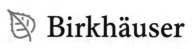

Frontiers in Mathematics

Advisory Editorial Board

More information about this series at http://www.springer.com/series/5388

José M. Mazón • Julio Daniel Rossi •
J. Julián Toledo

Nonlocal Perimeter, Curvature and Minimal Surfaces for Measurable Sets

 Birkhäuser

José M. Mazón
Departamento de Análisis Matemático
Universitat de València
Valencia, Spain

J. Julián Toledo
Departamento de Análisis Matemático
Universitat de València
València, Spain

Julio Daniel Rossi
Departamento de Matemáticas
Universidad de Buenos Aires
Buenos Aires, Argentina

ISSN 1660-8046   ISSN 1660-8054   (electronic)
Frontiers in Mathematics
ISBN 978-3-030-06242-2   ISBN 978-3-030-06243-9   (eBook)
https://doi.org/10.1007/978-3-030-06243-9

Library of Congress Control Number: 2018966514

Mathematics Subject Classification (2010): 45C99, 28A75, 49Q05, 45M05, 35K08

This book is published under the imprint Birkhäuser, www.birkhauser-science.com by the registered company Springer Nature Switzerland AG
The registered company address is: Gewerbestrasse 11, 6330 Cham, Switzerland

J. M. Mazón dedicates this book to Claudia

J. D. Rossi dedicates this book to Cecilia

J. Toledo dedicates this book to his parents, Julián and Feliciana

Preface

The goal in this monograph is to present recent results concerning nonlocal perimeter, curvature, and minimal surface that can be used for measurable sets. The concept of perimeter and curvature of planar forms goes back (at least) to the ancient Greece where mathematicians developed various geometric tools and ideas to try to give sense to different measures (area, length, etc.). In this context, the perimeter of a planar set was defined as the length of the curve that surrounds the set. Nowadays, there are modern concepts of perimeter and curvature that extend this intuitive idea and make it applicable to a wider class of sets. This modern point of view starts with Caccioppoli's seminal works and continued with the ideas developed by De Giorgi in the 1950s. The usual concepts of perimeter, curvature, and minimal surfaces require some regularity of the involved sets. For instance, to define the perimeter in geometric measure theory one asks for its characteristic function to be in the space BV (bounded variation functions); see [6, 57]. A fundamental result by De Giorgi and Federer shows that if we use this definition the perimeter coincides with the $(N-1)$-dimensional Hausdorff measure of a certain subset of the topological boundary, so that consistency with the naïve idea of being the length of the boundary is guaranteed. With this notion, not every bounded measurable set has finite perimeter. In the same spirit, when one wants to compute the curvature of the boundary of a set using classical tools one asks for this boundary to be twice differentiable. Here we focus on a new version of perimeter and curvature that can be applied (and gives a finite number) to any measurable set with finite measure. We follow our recent results in [58, 59].

The main idea to obtain a notion of perimeter that can be used to a large class of sets is to realize that the usual notions of perimeter and curvature involve the computation of *local* quantities. Here we work with a *nonlocal* analogous to these quantities. In this direction, recently, as analogous objects to the standard heat equation, nonlocal diffusion problems were proposed, like

$$u_t(x, t) = J * u(x, t) - u(x, t) = \int_{\mathbb{R}^N} J(x - y)(u(y, t) - u(x, t)) \, dy,$$

with a convolution kernel J that is assumed to be continuous, radially symmetric and non-increasing, nonnegative, compactly supported verifying $\int J = 1$ (some of these conditions can be relaxed according to the result one wants to prove). This nonlocal equation shares many properties with its local counterpart, the heat equation; for example, bounded stationary solutions are constants, there is a maximum principle, and there is infinite speed of propagation (compactly supported nonnegative and nontrivial initial conditions give rise to solutions that are strictly positive everywhere). In addition, the decay rate as t goes to infinity for both problems is the same. However, there is a major difference: the heat equation has an instantaneous smoothing effect; solutions become C^∞ for every positive time even if the initial condition is not differentiable; on the other hand, solutions to the nonlocal evolution equation are as smooth as the initial data are (there is no smoothing effect in this case). This fact can be explained looking at the fundamental solution, that is smooth (a Gaussian) for the heat equation, but is a smooth function plus a delta for the nonlocal problem; see the recent book [12]. This lack of regularizing effect is closely related to the fact that the right-hand side of the equation, $J * u - u$, makes sense for nonsmooth functions u.

Concerning the perimeter of a set, as we have mentioned, a usual definition in geometric measure theory is to compute the total variation of the characteristic function of the set,

$$Per(E) = \int_{\mathbb{R}^N} |D\chi_E|.$$

Looking at this approach it seems reasonable to take the limit as $p \searrow 1$ in the p-Laplacian type energy of a function u to obtain

$$\lim_{p \searrow 1} \int_{\mathbb{R}^N} |Du|^p = \int_{\mathbb{R}^N} |Du|.$$

In the nonlocal setting, there is an analogous to the p-energy given by

$$\frac{1}{2p} \int_{\mathbb{R}^N} \int_{\mathbb{R}^N} J(x - y)|u(y) - u(x)|^p \, dy \, dx.$$

If we take the limit as $p \searrow 1$ in this expression, we formally arrive to

$$\frac{1}{2} \int_{\mathbb{R}^N} \int_{\mathbb{R}^N} J(x - y)|u(y) - u(x)| \, dy \, dx,$$

that is a nonlocal quantity analogous to the total variation. Now, if we evaluate this expression in the characteristic function of a set, $u = \chi_E$, we get

$$\frac{1}{2} \int_{\mathbb{R}^N} \int_{\mathbb{R}^N} J(x - y)|\chi_E(y) - \chi_E(x)| \, dx \, dy = \int_E \int_{\mathbb{R}^n \setminus E} J(x - y) \, dy \, dx.$$

Let $E \subset \mathbb{R}^N$ be a measurable set; the *nonlocal J-perimeter of E* is defined by

$$P_J(E) := \int_E \left(\int_{\mathbb{R}^N \setminus E} J(x-y)\, dy \right) dx.$$

Note that if $|E| < +\infty$, we have

$$P_J(E) = |E| - \int_E \int_E J(x-y)\, dy\, dx.$$

To simplify the exposition, we are confining ourselves to study subsets of the Euclidean space $\mathbb{R}^N$; however, a large part of the results presented here can be extended to metric spaces with a measure (in this more general case, we have to consider a kernel $J(x, y)$).

This definition of perimeter is nonlocal in the sense that it is determined by the behavior of E in a neighborhood of the boundary ∂E. The quantity $P_J(E)$ measures the interaction between points of E and E^c via the interaction density $J(x - y)$. For J compactly supported in a ball $\overline{B_r(0)}$, the interaction is possible when the points $x \in E$ and $y \in E^c := \mathbb{R}^N \setminus E$ are both close to the boundary ∂E:

$$P_J(E) = \int_{\{x \in E\,:\,d(x,E^c)<r\}} \int_{\{y \in E^c\,:\,d(y,E)<r\}} J(x-y)\,dy\,dx.$$

The concept of nonlocal perimeter was introduced in [17, 36] as was explicitly stated in [20]. For singular kernels of the form $\frac{1}{|x|^{N+s}}$, $0 < s < 1$, the nonlocal perimeter reappeared later in [29]. See also the pioneering works [73, 74], where some functionals of this type were analyzed in connection with fractal dimensions. The nonlocal s-perimeter of $E \subset \mathbb{R}^N$ is defined (formally) as

$$Per_s(E) := \int_E \int_{E^c} \frac{1}{|x-y|^{N+s}}\, dx\, dy.$$

The usual notion of perimeter is recovered by the limit

$$\lim_{s \to 1} (1-s) Per_s(E) = Per(E);$$

see [7, 20, 31, 36]. Recently, the s-perimeter has inspired literature in different directions, including s-minimal surfaces, see [18, 29–31, 34, 37, 48, 66, 68] and the references therein, or image processing, see for instance [22, 23, 50] and the references therein. We also refer to [8, 18, 30, 31, 48, 66, 68] and the references therein.

As happens for the usual perimeter, since the kernel is singular at the origin (and not integrable) to have a finite s-perimeter one needs some regularity on the boundary of the involved set E.

Our aim here is to deal with nonsingular kernels. Since the kernel is not singular, the concept of perimeter is well defined for every measurable set and is finite for every set with finite measure. This is our starting point in this book. As for the case of singular kernels, as a consequence of the results in [36], the usual notion of perimeter is recovered by a limit rescaling formula:

$$\lim_{\epsilon \downarrow 0} \frac{C_J}{\epsilon} P_{J_\epsilon}(E) = \lim_{\epsilon \downarrow 0} C_J \epsilon^{N-1} P_J \left(\frac{E}{\epsilon} \right) = \text{Per}(E),$$

where J_ϵ are the rescaled kernels

$$J_\epsilon(x) := \frac{1}{\epsilon^N} J \left(\frac{x}{\epsilon} \right),$$

and the constant C_J is given by

$$C_J = \frac{2}{\displaystyle\int_{\mathbb{R}^N} J(z)|z_N|dz}.$$

Concerning the curvature, we can look at the first variation of the perimeter with respect to deformations of the domain and look for the equation verified by minimal sets. Let us be more precise and consider Ω an open bounded subset of $\mathbb{R}^N$. The notion of nonlocal perimeter can be localized to a bounded open set $\Omega \subset \mathbb{R}^N$ by setting formally

$$P_J(E, \Omega) := \int_E \int_{E^c} J(x-y)dydx - \int_{E\setminus\Omega} \int_{(E\cup\Omega)^c} J(x-y)dydx.$$

We say that a measurable set $E \subset \mathbb{R}^N$ is *J-minimal in Ω* if

$$P_J(E, \Omega) \le P_J(F, \Omega) \text{ for any measurable set } F \text{ such that } F \setminus \Omega = E \setminus \Omega.$$

We will see that such minimal sets verify the equation

$$\int_{\mathbb{R}^N} J(x-y) \left(\chi_{E^c}(y) - \chi_E(y) \right) dy = 0,$$

for all $x \in \partial E \cap \Omega$ such that $|E \cap B_\delta(x)| > 0$ and $|E^c \cap B_\delta(x)| > 0$ for every small δ. So, by analogy with the classical case, we consider the left-hand side as a nonlocal mean curvature, denoted by

$$H_{\partial E}^J(x) := \int_{\mathbb{R}^N} J(x-y) \left(\chi_{E^c}(y) - \chi_E(y) \right) dy.$$

Note that this curvature $H_{\partial E}^J(x)$ makes sense for every $x \in \mathbb{R}^N$ and not only for points x on the boundary of the set E. For this nonlocal curvature, we can also show an approximation result to recover the usual notion of mean curvature. In fact, when $E \subset \mathbb{R}^N$ is a smooth set with C^2 boundary, for every $x \in \partial E$, we have

$$\lim_{\epsilon \downarrow 0} \frac{C_J}{\epsilon} H_{\partial E}^{J_\epsilon}(x) = (N - 1)H_{\partial E}(x),$$

where $H_{\partial E}(x)$ is the (local) mean curvature of ∂E at x.

Once we have a notion of curvature, one can study a mean curvature flow, that is, the evolution of a set when we prescribe that the normal velocity of a point on the boundary is proportional to the curvature of the set at that point. To study such a flow we need to introduce a normal vector defined for nonsmooth sets. After this is done we are able to show that, as happens for the local mean curvature flow, a ball evolves through this flow as a collection of balls that shrink to a point in finite time.

The study of nonlocal perimeters is motivated by both theory and applications as we will briefly describe below. Note that in applications, it is also natural to consider interactions that are not homogeneous or rotationally invariant (but here we restrict ourselves to this class of kernels to simplify the exposition).

The first application that we present is related to image processing and bitmaps; see [33]. Let us consider the framework of BMP-type images with square pixels of (small) size $\delta > 0$ (suppose that $1/\delta \in \mathbb{N}$ for simplicity). As an example, let us consider a picture of a square of side of length one, with sides at 45 degrees with respect to the orientation of the pixels. Take the square with vertices at $(1/\sqrt{2}, 0)$, $(0, 1/\sqrt{2})$, $(-1/\sqrt{2}, 0)$, $(0, -1/\sqrt{2})$ and pixels of the form $(k\delta, (k + 1)\delta) \times (j\delta, (j + 1)\delta)$. Let us look at the "computational image" composed by the pixels (small squares whose sides are of length δ) that intersects the square and let us compare it with the original square. Note that the "discretization" of the image is composed by a large number of pixels. In this configuration, the classical perimeter functional provides a rather inaccurate tool to analyze the perimeter of the original square, no matter how small the pixels are, i.e., no matter how good is the image resolution. Indeed, the perimeter of the ideal square is 4, while the perimeter of the picture displayed by the monitor (the one composed by the pixels) is always $4\sqrt{2}$ (independently of the size of δ). Hence, the classical perimeter is always producing an error by a factor $\sqrt{2}$, even in cases of extremely high resolution. Instead, the nonlocal perimeter studied here or other nonlocal perimeters would provide a much better approximation of the classical perimeter of the ideal square in the case of high image resolution. Indeed, note that both the nonlocal perimeter and the nonlocal mean curvature at a point x are continuous as functions of E in the following sense: if $E_n \to E$ in the sense that $|E_n \triangle E| \to 0$ then

$$P_J(E_n) \to P_J(E) \qquad \text{and} \qquad H_{\partial E_n}^J(x) \to H_{\partial E}^J(x).$$

Hence, as the resolution of the image gets better (δ becomes very small) the measure of the difference of the sets (the real set vs the union of the pixels) goes to zero (the symmetric difference has measure of order 4δ) and then the nonlocal perimeter of the computational image gets very close to the nonlocal perimeter of the original square. Therefore, in this case, the nonlocal perimeter provides more precise information than the classical one.

For applications of nonlocal operator to image processing, see for instance [13, 14, 41, 50, 53].

Another motivation for the study of nonlocal minimal surfaces comes from models describing phase-transitions problems with long-range interactions; see [25–28, 67]. In these models, the minimal surface appears in a limit in which the zone that connects the two stable states concentrates along a surface.

Let us now summarize the contents of this book. After a brief section in which we fix the notation that will be used along the whole book, the content of the different chapters can be described as follows:

- In Chap. 1, we introduce the nonlocal perimeter, compare it with its local counterpart, and show that the local perimeter can be recovered from the nonlocal one by rescaling the kernel.
- In Chap. 2, we analyze some properties of the nonlocal perimeter and prove a nonlocal isoperimetric inequality and a nonlocal co-area formula.
- In Chap. 3, we introduce nonlocal minimal surfaces and the nonlocal curvature. Here we also include the proof that the local curvature can be obtained as a limit of the nonlocal curvature when the kernel is rescaled.
- Chapter 4, is devoted to develop a nonlocal functional setting that involves the nonlocal analogous to bounded variation functions and its relation with the nonlocal perimeter.
- In Chap. 5, we deal with nonlocal Cheeger and calibrable sets.
- In Chap. 6, we describe the nonlocal heat content of a set and study its behavior for small t. In this description, the nonlocal perimeter plays a crucial role.
- Finally, in Chap. 7 we introduce a nonlocal normal vector and, associated with this normal vector, we state a nonlocal mean curvature flow problem.

The bibliography of this monograph does not escape the usual rule of being incomplete. In general, we have listed those papers which are closer to the topics discussed here. But, even for those papers, the list is far from being exhaustive and we apologize for any possible omission.

Valencia, Spain   José M. Mazón
Buenos Aires, Argentina   Julio Daniel Rossi
Valéncia, Spain   J. Julián Toledo
October 2018

Acknowledgment

The authors have been partially supported by the Spanish MEC and FEDER, project MTM2015-70227-P.

Notations

The ambient space will be the Euclidean space $\mathbb{R}^N$. Throughout the book we will use the following notations:

$B_r(x)$ denotes the ball centered at $x \in \mathbb{R}^N$ of radius $r > 0$.
B_r denotes the ball centered at $0 \in \mathbb{R}^N$ of radius $r > 0$.
$\overline{B_r(x)}$ and $\overline{B_r}$ denote the closure of such balls.
The measure of B_1 is denoted by ω_N:

$$\omega_N = |B_1|.$$

The Hausdorff $(N-1)$-dimensional measure of a set $A \subset \mathbb{R}^N$ is denoted by $\mathcal{H}^{N-1}(A)$. The boundary of ∂B_1 is denoted by S^{N-1}, for which

$$\mathcal{H}^{N-1}(S^{N-1}) = N\omega_N.$$

We will write $(r, \overline{0})$ to denote the point $(r, 0, 0, \ldots, 0) \in \mathbb{R}^N, r \in \mathbb{R}$.

Contents

Nonlocal Perimeter

1.1   The Classical Concept of Perimeter

The word *perimeter* comes from the Greek *peri* (around) and *meter* (measure). A perimeter is usually used with two senses: it is *the boundary* that surrounds an N-dimensional set, and it is *the measure of such boundary*. We will see in these two first sections that these two concepts must be well precise. Nevertheless, for a set $E \subset \mathbb{R}^N$ with smooth C^2 boundary its perimeter is clearly defined as

$$\mathrm{Per}(E) = \mathcal{H}^{N-1}(\partial E),$$

being $\mathcal{H}^{N-1}$ the Hausdorff $(N-1)$-dimensional measure.

The study of sets of finite perimeter for general sets without smooth boundary goes back to Renato Caccioppoli, who in 1927 defined the measure of the boundary of an open set of the plane using the total variation in the sense of Tonelli. More precisely he says that E is a set of finite perimeter if there exist polyhedral sets E_n such that $E_n \to E$ locally in measure and

$$\sup_{n \in \mathbb{N}} \mathrm{Area}(\partial E_n) < +\infty.$$

In 1951 Caccioppoli, by using the generalization of functions of bounded variation to the case of several variables given by Lamberto Cesari in 1936, suggested to study the

© Springer Nature Switzerland AG 2019   1
J. M. Mazón et al., *Nonlocal Perimeter, Curvature and Minimal Surfaces for Measurable Sets*, Frontiers in Mathematics,
https://doi.org/10.1007/978-3-030-06243-9_1

geometrical properties of the Lebesgue measurable sets E with

$$\int_E \text{div}\phi(x)dx \leq K\|\phi\|_\infty, \quad \forall \phi \in C_0^1(\mathbb{R}^N, \mathbb{R}^N), \tag{1.1}$$

whenever a positive real number K exists.

Caccioppoli was aware that (1.1) is equivalent to the existence of a vector finite Radon measure μ_E with

$$\int_E \text{div}\phi(x)dx = \int_{\partial E} \phi \cdot d\mu_E, \quad \forall \phi \in C_0^1(\mathbb{R}^N, \mathbb{R}^N), \tag{1.2}$$

and that the total variation $|\mu_E|$ of μ_E satisfies

$$|\mu_E|(\mathbb{R}^N) \leq K.$$

In 1952, at the Salzburg Congress of the Austrian Mathematical Society, Ennio De Giorgi presented his first results on this subject developing the ideas of Caccioppoli.

De Giorgi original definition of perimeter of a measurable subset $E \subset \mathbb{R}^N$ was based on the heat semigroup in $\mathbb{R}^N$, because of its regularizing effect, and can be described as follows:

$$\text{Per}(E) := \lim_{t \to 0} \|\nabla_x T(t)\chi_E\|_{L^1(\mathbb{R}^N)},$$

where χ_E denotes the characteristic function of E, and $(T(t))_{t \geq 0}$ is the heat semigroup, that is, if $p_N : \mathbb{R}^N \times \mathbb{R}^n \times \mathbb{R} \to \mathbb{R}$ is the Gauss–Weierstrass kernel, defined by

$$p_N(x, y, t) = \frac{1}{(4\pi t)^{\frac{N}{2}}} e^{-\frac{|x-y|^2}{4t}},$$

then

$$T(t)u(x) = \int_{\mathbb{R}^N} p_N(x, y, t)u(y)dy,$$

for every $u \in L^1(\mathbb{R}^N)$; and he proved that, whenever μ_E is the Radon vector measure satisfying (1.2),

$$\text{Per}(E) = |\mu_E|(\mathbb{R}^N).$$

We refer also to Chap. 6 for another look of this interplay with the heat evolution problem in the nonlocal perimeter framework.

He also announced the validity of the isoperimetrical inequality for Lebesgue measurable sets, that is, the inequality

$$\min\left\{|E|, |\mathbb{R}^N \setminus E|\right\} \leq \left(\mathrm{Per}(E)\right)^{\frac{N}{N-1}} \tag{1.3}$$

for any measurable set E and any $N \geq 2$, and conjectured that

$$\inf\left\{\mathrm{Per}(E) \; : \; |E| = 1\right\} = \mathrm{Per}(B)$$

for B the ball of measure one. The proofs of all these results appear in [38] and [39]. After the death of Cacciopoli, De Giorgi started to call the sets of finite perimeter as Caccioppoli sets.

In 1960 Federer and Fleming [43] introduced a new approach to the theory by mean of the normal and integral currents. In his famous book on Geometric Measure Theory [42], Federer showed that Caccioppoli sets are normal currents of dimension N in N-dimensional Euclidean spaces.

Now, after the paper of Miranda [60], the easy and more traditional way to study the theory of sets of finite perimeter is the approach that uses functions of bounded variation in the framework of the theory of distributions. Two excellent monographs about this approach are [6, 51]. Let us overview this approach.

Let Ω be an open bounded subset of $\mathbb{R}^N$. A function $u \in L^1(\Omega)$ whose partial derivatives in the sense of distributions are measures with finite total variation in Ω is called a function of bounded variation. The class of such functions will be denoted by $BV(\Omega)$ (see [6, 51]):

$$BV(\Omega) = \{u \in L^1(\Omega) \; : \; |Du|(\Omega) < \infty\},$$

where $|Du|$ is the total variation of the distributional gradient of u, which turns out to be

$$|Du|(\Omega) = \sup\left\{\int_\Omega u \operatorname{div}\phi\, dx : \phi \in C_0^\infty(\Omega, \mathbb{R}^N), |\phi(x)| \leq 1 \text{ for } x \in \Omega\right\},$$

and will be also denoted by $\int_\Omega |Du|$.

For $u \in BV(\Omega)$, the gradient Du is a Radon measure that decomposes into its absolutely continuous and singular parts $Du = D^a u + D^s u$. Then $D^a u = \nabla u\, \mathcal{L}^N$, where ∇u is the Radon–Nikodym derivative of the measure Du with respect to the Lebesgue measure $\mathcal{L}^N$.

A measurable set $E \subset \mathbb{R}^N$ is said to be of *finite perimeter in* Ω if the characteristic function χ_E of E belongs to $BV(\Omega)$, in this case the perimeter of E is defined as

$$\mathrm{Per}(E, \Omega) = |D\chi_E|(\Omega).$$

A measurable set $E \subset \mathbb{R}^N$ is said to be of *finite perimeter* if $\mathcal{X}_E \in BV(\mathbb{R}^N)$, and we denote

$$\mathrm{Per}(E) = \mathrm{Per}(E, \mathbb{R}^N).$$

Then, we have that

$$\mathrm{Per}(E) = \sup \left\{ \int_{\mathbb{R}^N} \mathcal{X}_E(x) \mathrm{div}\phi(x) dx \; : \; \phi \in C_c^\infty(\mathbb{R}^N, \mathbb{R}^N) \; \|\phi\|_\infty \le 1 \right\}. \qquad (1.4)$$

If $u \in BV(\mathbb{R}^N)$, the energy functional associated with the total variation is

$$\mathcal{TV}(u) = \int_{\mathbb{R}^N} |Du|, \qquad (1.5)$$

that is,

$$\mathcal{TV}(u) = \sup \left\{ \int_{\mathbb{R}^N} u(x) \mathrm{div}\phi(x) dx \; : \; \phi \in C_c^\infty(\mathbb{R}^N, \mathbb{R}^N) \; \|\phi\|_\infty \le 1 \right\}.$$

Then, for a measurable set $E \subset \mathbb{R}^N$ having finite perimeter,

$$\mathrm{Per}(E) = \int_{\mathbb{R}^N} |D\mathcal{X}_E|. \qquad (1.6)$$

If $E \subset \mathbb{R}^N$ is a bounded open set with boundary ∂E of class C^2, then E is a set of finite perimeter and

$$\mathrm{Per}(E) = \mathcal{H}^{N-1}(\partial E). \qquad (1.7)$$

In fact, if $\phi \in C_c^\infty(\mathbb{R}^N, \mathbb{R}^N)$ with $\|\phi\|_\infty \le 1$, applying the Gauss–Green formula, we have

$$\int_E \mathrm{div}\phi(x) dx = \int_{\partial E} \phi \cdot \nu d\mathcal{H}^{N-1} \le \mathcal{H}^{N-1}(\partial E),$$

where ν is the unit outward normal ν to ∂E. Then, by (1.4), we have

$$\mathrm{Per}(E) \le \mathcal{H}^{N-1}(\partial E).$$

Let us see the another inequality. Since ∂E is of class C^2, there exists an open set U containing ∂E, such that $d(x) := \mathrm{dist}(x, E)$ is of class C^1 on $U \setminus \partial E$ and

$$\nabla d(x) = \frac{x - \xi(x)}{d(x)},$$

being $\xi(x)$ the unique point of ∂E such that $|x - \xi(x)| = d(x)$. Hence, the unit outward normal ν to ∂E has an extension $\hat{\nu} \in C_c^1(\mathbb{R}^N)$ such that $|\hat{\nu}| \leq 1$. Therefore, if we take $\phi = \eta \hat{\nu}$, with $\eta \in C_c^\infty(\mathbb{R}^N)$, we get

$$\int_E \text{div}\phi(x)dx = \int_{\partial E} \eta d\mathcal{H}^{N-1}.$$

Therefore

$$\text{Per}(E) \geq \sup\left\{\int_{\partial E} \eta \, d\mathcal{H}^{N-1} \; : \; \eta \in C_c^\infty(\mathbb{R}^N), \; |\eta| \leq 1\right\} = \mathcal{H}^{N-1}(\partial E),$$

and we finish the proof of (1.7).

In the case that the set E has no smooth boundary, the topological boundary ∂E of a set of finite perimeter is not a good candidate to measure the perimeter because its Hausdorff measure exceeds, in general, such value. The correct boundary in this context is the *reduced boundary* introduced by De Giorgi. The *reduced boundary* of a set of finite perimeter $E \subset \mathbb{R}^N$, denoted by $\mathcal{F}E$, is defined as the collection of points x at which the limit

$$\nu_E(x) := \lim_{\rho \downarrow 0} \frac{D\chi_E(B_\rho(x))}{|D\chi_E|(B_\rho(x))}$$

exists and has length equal to one, i.e.,

$$|\nu_E(x)| = 1.$$

The function $\nu_E : \mathcal{F}E \to S^{N-1}$ is called the *generalized inner normal* to E.

De Giorgi proved the following generalization of the Gauss–Green formula (see [6, Theorem 3.36] for a proof):

$$\int_E \text{div}\phi(x)dx = -\int_\Omega \nu_E \cdot \phi \, d|D\chi_E| \quad \forall \phi \in C_c^1(\Omega, \mathbb{R}^N),$$

for any set E of finite perimeter in Ω, where $D\chi_E = \nu_E|D\chi_E|$ is the polar decomposition of $D\chi_E$. And he also proved that

$$\text{Per}(E) = \mathcal{H}^{N-1}(\mathcal{F}E),$$

which is a generalization of the classical formula (1.7) (see [6, Theorem 3.59] for a proof). This result provides geometric intuition for the notion of reduced boundary and confirms that the above definition of sets of finite perimeter is the more natural.

In 1960 Fleming and Rishel [46] proved the following *coarea formula* (see [6, Theorem 3.40] for a proof). Let $\Omega \subset \mathbb{R}^N$ be an open set. If $u \in BV(\Omega)$, then the set $\{u > t\}$ has finite perimeter in Ω for $\mathcal{L}^1$-a.e. $t \in \mathbb{R}$. Moreover,

$$|Du|(B) = \int_{-\infty}^{+\infty} |D\chi_{\{u>t\}}|(B)dt \tag{1.8}$$

and

$$Du(B) = \int_{-\infty}^{+\infty} D\chi_{\{u>t\}}(B)dt,$$

for any Borel set $B \subset \Omega$.

As mentioned in the introduction, the concept of nonlocal perimeter was introduced in [17, 36], and for kernels of the form $\frac{1}{|x|^{N+s}}$, $0 < s < 1$, the nonlocal perimeter reappeared in [29]. The s-perimeter of $E \subset \mathbb{R}^N$ is defined formally as

$$\mathrm{Per}_s(E) = \int_E \int_{\mathbb{R}^N \setminus E} \frac{1}{|x-y|^{N+s}} dx dy,$$

and, if $\mathrm{Per}_s(E) < \infty$, it can be rewritten as

$$\mathrm{Per}_s(E) = \frac{1}{2} \int_{\mathbb{R}^N} \int_{\mathbb{R}^N} \frac{|\chi_E(y) - \chi_E(x)|}{|x-y|^{N+s}} dx dy.$$

1.2   Nonlocal Perimeter for Non-singular Kernels

Let $J : \mathbb{R}^N \to [0, +\infty]$ be a radially symmetric **non-singular kernel**, that is, a measurable, nonnegative, and radially symmetric function verifying

$$\int_{\mathbb{R}^N} J(z)dz = 1.$$

This conditions on J will be considered all around the book, but we will impose other conditions as continuity, radially non-increasing, etc., that will be given when necessary.

Associated with this kernel J, a nonlocal perimeter is defined for general measurable sets. Let $E \subset \mathbb{R}^N$ be a measurable set, the *nonlocal J-perimeter of E* is defined by

$$P_J(E) = \int_E \left(\int_{\mathbb{R}^N \setminus E} J(x-y)dy \right) dx. \tag{1.9}$$

Implicitly when the integral is not finite, we are setting $P_J(E) = +\infty$. We have that

$$P_J(E) = P_J(\mathbb{R}^N \setminus E).$$

It is easy to see that

$$P_J(E) = \frac{1}{2} \int_{\mathbb{R}^N} \int_{\mathbb{R}^N} J(x - y)|\chi_E(y) - \chi_E(x)|dxdy, \qquad (1.10)$$

and that, if $|E| < +\infty$,

$$P_J(E) = |E| - \int_E \int_E J(x - y)dydx. \qquad (1.11)$$

Observe that (1.11) is not true in general for the local case where bounded sets can have infinite perimeter.

This definition of perimeter is nonlocal in the sense that it is determined by the behavior of E in a neighborhood of the boundary ∂E. The quantity $P_J(E)$ measures the interaction between points of E and $\mathbb{R}^N \setminus E$ via the interaction density $J(x - y)$. For J compactly supported in the ball $\overline{B_r}$, the interaction is possible when the points $x \in E$ and $y \in \mathbb{R}^N \setminus E$ are both close to the boundary ∂E:

$$P_J(E) = \int_{\{x \in E : d(x, \mathbb{R}^N \setminus E) < r\}} \int_{\{y \in \mathbb{R}^N \setminus E : d(y, E) < r\}} J(x - y)dydx.$$

Since the kernel J is not singular, the concept of perimeter is well defined for every measurable set and is finite for every set with finite measure. This concept does not depend on any regularity condition.

The notion of nonlocal perimeter can be localized to a bounded open set $\Omega \subset \mathbb{R}^N$ by setting formally (see (1.13))

$$P_J(E, \Omega) = \int_E \int_{\mathbb{R}^N \setminus E} J(x - y)dydx - \int_{E \setminus \Omega} \int_{\mathbb{R}^N \setminus (E \cup \Omega)} J(x - y)dydx. \qquad (1.12)$$

Associated with J there is a nonlocal interaction between two measurable sets A and B in $\mathbb{R}^N$:

$$L_J(A, B) = \int_A \int_B J(x - y)dydx.$$

We have that

$$L_J(A, B) = L_J(B, A).$$

With this concept at hand we can rewrite the J-perimeter of the measurable set E as

$$P_J(E) = L_J(E, \mathbb{R}^N \setminus E),$$

and the localized nonlocal perimeter is defined as

$$P_J(E, \Omega) = L_J(E \cap \Omega, \mathbb{R}^N \setminus E) + L_J(E \setminus \Omega, \Omega \setminus E). \tag{1.13}$$

Observe that

$$L_J(E, \mathbb{R}^N \setminus E) = L_J(E \cap \Omega, \mathbb{R}^N \setminus E) + L_J(E \setminus \Omega, \Omega \setminus E) + L_J(E \setminus \Omega, \mathbb{R}^N \setminus (E \cup \Omega)),$$

from where we get the formal definition given in (1.12), where $+\infty - (+\infty)$ could be possible.

Observe that, if we assume that in B (a measurable set), there is an homogeneous population (we assume density 1 of the population inside B) and $J(x - y)$ is the probability that an individual jumps from y to x, then $L_J(A, B)$ is the total amount of individuals that arrives to A from B. Note that one can think all this the other way around, if we assume that in A there is a population with density 1, then $L_J(A, B)$ is the amount of individuals that goes from A to B. Then we can remark that the individuals that go from $\mathbb{R}^N \setminus E$ to E travel across the boundary of the set E, and, therefore, (1.9) is a natural way of defining the perimeter of E; we are counting the total flux of individuals that crosses the boundary when they go from $\mathbb{R}^N \setminus E$ to E.

Let us now give some properties of the nonlocal J-perimeter. The first ones are easy to obtain.

Proposition 1.1 *It holds that, for all set E of Lebesgue finite measure,*

$$P_J(E) \le |E|, \tag{1.14}$$

$$P_J(E) = P_J(z + E) \quad \text{for all } z \in \mathbb{R}^N,$$

and

$$P_J(E) = P_J(rot_\alpha(E)) \quad \text{for all rotation } rot_\alpha.$$

Proposition 1.2 *Let A, $B \subset \mathbb{R}^N$ be the measurable sets with $A \cap B = \emptyset$. Then,*

$$P_J(A \cup B) = P_J(A) + P_J(B) - 2L_J(A, B).$$

In particular, if J is compactly supported and

$$d(A; B) := \inf\left\{ |x - y| \; : \; x \in A, \; y \in B \right\} > \frac{1}{2} diam(supp(J)),$$

then

$$P_J(A \cup B) = P_J(A) + P_J(B).$$

Proof We have

$$
\begin{aligned}
P_J(A \cup B) &= \int_{A \cup B} \left(\int_{\mathbb{R}^N \setminus (A \cup B)} J(x - y) dy \right) dx \\
&= \int_A \left(\int_{\mathbb{R}^N \setminus (A \cup B)} J(x - y) dy \right) dx + \int_B \left(\int_{\mathbb{R}^N \setminus (A \cup B)} J(x - y) dy \right) dx \\
&= \int_A \left(\int_{\mathbb{R}^N \setminus A} J(x - y) dy - \int_B J(x - y) dy \right) dx \\
&\quad + \int_B \left(\int_{\mathbb{R}^N \setminus B} J(x - y) dy - \int_A J(x - y) dy \right) dx \\
&= P_J(A) + P_J(B) - 2 \int_A \left(\int_B J(x - y) dy \right) dx,
\end{aligned}
$$

as we wanted to show. □

Corollary 1.3 *Let A, B, C be the pairwise disjoints measurable sets in $\mathbb{R}^N$. Then*

$$P_J(A \cup B \cup C) = P_J(A \cup B) + P_J(A \cup C) + P_J(B \cup C)$$

$$- P_J(A) - P_J(B) - P_J(C).$$

Proof By Proposition 1.2 we have

$$P_J(A \cup B \cup C) = P_J(A \cup B) + P_J(C) - 2L_J(A \cup B, C). \tag{1.15}$$

Now, again by Proposition 1.2,

$$2L_J(A \cup B, C) = 2L_J(A, C) + 2L_J(B, C)$$

$$= -P_J(A \cup C) + P_J(A) + P_J(C) \tag{1.16}$$

$$- P_J(B \cup C) + P_J(B) + P_J(C).$$

Hence, putting (1.16) in (1.15) we obtain the desired conclusion. □

Associated with a kernel J we define the space

$$BV_J(\mathbb{R}^N) = \left\{ u : \mathbb{R}^N \to \mathbb{R} \text{ measurable} : \right.$$

$$\left. \int_{\mathbb{R}^N} \int_{\mathbb{R}^N} J(x - y)|u(y) - u(x)|dxdy < \infty \right\}.$$

We have that $L^1(\mathbb{R}^N) \subset BV_J(\mathbb{R}^N)$. For $u \in BV_J(\mathbb{R}^N)$, we also define the functional

$$\mathcal{TV}_J(u) = \frac{1}{2} \int_{\mathbb{R}^N} \int_{\mathbb{R}^N} J(x - y)|u(y) - u(x)|dxdy.$$

By (1.10) we have

$$P_J(E) = \mathcal{TV}_J(\mathcal{X}_E). \tag{1.17}$$

The functional $\mathcal{TV}_J$ is the nonlocal analog of the energy functional associated with the total variation (1.5). The nonlocal perimeter $P_J(E)$, written in the form (1.17), can be seen as the nonlocal version of $\text{Per}(E)$ given in (1.6).

For $\Omega \subset \mathbb{R}^N$ measurable, we can also define the total J-variation of u in Ω as

$$\mathcal{TV}_J(u, \Omega) = \frac{1}{2} \int_\Omega \int_\Omega J(x - y)|u(y) - u(x)|dydx$$

$$+ \frac{1}{2} \int_\Omega \int_{\mathbb{R}^N \setminus \Omega} J(x - y)|u(y) - u(x)|dydx$$

$$+ \frac{1}{2} \int_{\mathbb{R}^N \setminus \Omega} \int_\Omega J(x - y)|u(y) - u(x)|dydx.$$

We have that

$$\mathcal{TV}_J(u, \Omega) = \frac{1}{2} \int_\Omega \int_\Omega J(x - y)|u(y) - u(x)|dydx$$

$$+ \int_\Omega \int_{\mathbb{R}^N \setminus \Omega} J(x - y)|u(y) - u(x)|dydx$$

$$= \frac{1}{2} \int_{\mathbb{R}^N \times \mathbb{R}^N \setminus (\mathbb{R}^N \setminus \Omega) \times (\mathbb{R}^N \setminus \Omega)} J(x - y)|u(y) - u(x)|dxdy.$$

We have

$$\mathcal{TV}_J(u, \mathbb{R}^N) = \mathcal{TV}_J(u).$$

With the above functional we can prove that

$$P_J(E, \Omega) = \mathcal{TV}_J(\chi_E, \Omega). \tag{1.18}$$

Indeed,

$$
\mathcal{TV}_J(\chi_E, \Omega) = \frac{1}{2} \int_\Omega \int_\Omega J(x - y) |\chi_E(y) - \chi_E(x)| \, dy \, dx
$$

$$
+ \int_\Omega \int_{\mathbb{R}^N \setminus \Omega} J(x - y) |\chi_E(y) - \chi_E(x)| \, dy \, dx
$$

$$
= \frac{1}{2} \int_\Omega \int_\Omega J(x - y) |\chi_E(y) - \chi_E(x)| \, dy \, dx
$$

$$
+ \int_\Omega \int_{\mathbb{R}^N \setminus \Omega} J(x - y) (\chi_E(x) - \chi_E(y))^+ \, dy \, dx
$$

$$
+ \int_\Omega \int_{\mathbb{R}^N \setminus \Omega} J(x - y) (\chi_E(x) - \chi_E(y))^- \, dy \, dx.
$$

Now, it is easy to see that

$$
\frac{1}{2} \int_\Omega \int_\Omega J(x - y) |\chi_E(y) - \chi_E(x)| \, dy \, dx = L_J(\Omega \cap E, \Omega \setminus E),
$$

$$
\int_\Omega \int_{\mathbb{R}^N \setminus \Omega} J(x - y) (\chi_E(x) - \chi_E(y))^+ \, dy \, dx = L_J(\Omega \cap E, \mathbb{R}^N \setminus (\Omega \cup E)),
$$

and

$$
\int_\Omega \int_{\mathbb{R}^N \setminus \Omega} J(x - y) (\chi_E(x) - \chi_E(y))^- \, dy \, dx = L_J(E \setminus \Omega, \Omega \setminus E).
$$

Then, since

$$
P_J(E, \Omega) = L_J(\Omega \cap E, \Omega \setminus E) + L_J(\Omega \cap E, \mathbb{R}^N \setminus (\Omega \cup E)) + L_J(E \setminus \Omega, \Omega \setminus E),
$$

we conclude that (1.18) holds true.

We have the following relation between the local and the nonlocal J-perimeter.

Proposition 1.4 *Assume that J satisfies*

$$
M_J := \int_{\mathbb{R}^N} J(z) |z| \, dz < +\infty. \tag{1.19}
$$

For every $u \in BV(\mathbb{R}^N)$ we have

$$\mathcal{TV}_J(u) \leq \frac{M_J}{2} \mathcal{TV}(u). \tag{1.20}$$

In particular, for every set of finite perimeter $E \subset \mathbb{R}^N$,

$$P_J(E) \leq \frac{M_J}{2} \mathrm{Per}(E). \tag{1.21}$$

Proof Given $u \in BV(\mathbb{R}^N)$ there exists a sequence $\{u_n\}_{n \in \mathbb{N}} \subset C^\infty(\mathbb{R}^N) \cap BV(\mathbb{R}^N)$ such that

$$\lim_{n \to \infty} \|u_n - u\|_{L^1(\mathbb{R}^N)} = 0$$

and

$$\lim_{n \to \infty} \int_{\mathbb{R}^N} |\nabla u_n| dx = \int_{\mathbb{R}^N} |Du|.$$

Then, in order to prove (1.20) it will be sufficient to prove it for $u \in C^\infty(\mathbb{R}^N) \cap BV(\mathbb{R}^N)$.
    Let us write

$$\mathcal{TV}_J(u) = \frac{1}{2} \int_{\mathbb{R}^N} J(z) \int_{\mathbb{R}^N} |u(x+z) - u(x)| \, dx dz.$$

Observe that we have

$$|u(x+z) - u(x)| \leq \left(\int_0^1 |\nabla u(x+tz)| dt \right) |z|.$$

Then,

$$\mathcal{TV}_J(u) \leq \frac{1}{2} \int_{\mathbb{R}^N} J(z) \int_{\mathbb{R}^N} \left(\int_0^1 |\nabla u(x+tz)| dt \right) |z| \, dx dz$$

$$= \frac{1}{2} \int_{\mathbb{R}^N} J(z)|z| \int_0^1 \left(\int_{\mathbb{R}^N} |\nabla u(x+tz)| dx \right) dt dz$$

$$= \frac{M_J}{2} \int_{\mathbb{R}^N} |\nabla u(\xi)| d\xi,$$

and we have proved (1.20) for smooth functions.   □

Observe that if J has compact support, then J satisfies (1.19).

Remark 1.5 For $r > 0$ consider the rescaled kernel

$$J_r(x) = \frac{1}{r^N} J\left(\frac{x}{r}\right).$$

Observe that J_r has the same total mass as J, and verifies

$$P_{J_r}(E) = r^N P_J\left(\frac{1}{r}E\right), \tag{1.22}$$

for all set E of Lebesgue finite measure. Observe also that

$$M_{J_r} = r M_J.$$

Then, under condition (1.19), (1.21) implies that

$$P_{J_r}(E) \le \frac{M_{J_r}}{2} \operatorname{Per}(E) = r \frac{M_J}{2} \operatorname{Per}(E), \tag{1.23}$$

for E with Lebesgue finite measure.

Observe that, from (1.23), if r converges to 0, then $P_{J_r}(E)$ also does it, nevertheless we will see in Theorem 1.11 that, for bounded sets of finite perimeter,

$$\lim_{r \to 0} \frac{1}{r} P_{J_r}(E) = c \operatorname{Per}(E),$$

for an adequate constant $c > 0$.

1.3   Rescaling the Nonlocal Perimeter

Assume that J satisfies (1.19). Consider the rescaled kernel introduced in Remark 1.5,

$$J_\epsilon(x) = \frac{1}{\epsilon^N} J\left(\frac{x}{\epsilon}\right), \quad \epsilon > 0.$$

Let also consider

$$C_J = \frac{2}{\displaystyle\int_{\mathbb{R}^N} J(z)|z_N|dz}.$$

Observe that

$$C_{J_\epsilon} = \frac{1}{\epsilon} C_J.$$

In [12] it is proved that the solutions u_ϵ of equations of the form

$$u_t(x,t) = C_{J_\epsilon} \int J_\epsilon(x-y) \frac{u(y,t)-u(x,t)}{|u(y,t)-u(x,t)|} dy$$

converge (up to a subsequence) to the solution of

$$u_t = \Delta_1 u,$$

for different boundary conditions, being $\Delta_1 u = \mathrm{div}\left(\frac{Du}{|Du|}\right)$.

Our aim in this section is to study which is the behavior of the nonlocal J-perimeter under rescaling.

Example 1.6 Take $J = \frac{1}{2}\chi_{[-1,1]}$. Then, a simple calculation gives

$$P_J([a,b]) = \frac{1}{2}\left(1 - \left(\left(1-(b-a)\right)^+\right)^2\right), \tag{1.24}$$

that is,

$$P_J([a,b]) = \begin{cases} \dfrac{1}{2} & \text{if } b-a > 1, \\[2ex] (b-a)\left(1 - \dfrac{1}{2}(b-a)\right) & \text{if } b-a \le 1. \end{cases}$$

Let us compute $\lim_{\epsilon \to 0} \frac{C_J}{\epsilon} P_{J_\epsilon}(E)$. Observe that for this particular kernel, we have $C_J = 4$. On account of (1.22) and (1.24), for ϵ small, we have

$$\frac{4}{\epsilon} P_{J_\epsilon}([a,b]) = 4 P_J\left(\left[\frac{a}{\epsilon}, \frac{b}{\epsilon}\right]\right) 2 = \mathrm{Per}([a,b]).$$

Example 1.7 Let us now consider $J(x) = w(\|x\|)\chi_{B_R}(x)$, with $w(r) = C_R r^\alpha$. Then

$$\int_{\mathbb{R}^N} J(x)dx = C_R \int_{B_R} \|x\|^\alpha dx = C_R \int_0^R \left(\int_{\partial B_r} r^\alpha d\sigma\right) dr$$

$$= C_R N\omega_N \int_0^R r^{\alpha+N-1} dr = \frac{C_R N\omega_N}{N+\alpha} R^{N+\alpha} = C_R \frac{N|B_R|R^\alpha}{N+\alpha}.$$

Then, if $\alpha > -N$ and $C_R = \frac{N+\alpha}{N|B_R|R^\alpha}$, we have $\int_{\mathbb{R}^N} J(x)dx = 1$. Consider the case $N = 1$, $R = 1$, and $\alpha = -\frac{1}{2}$, which corresponds to

$$J(x) = \frac{1}{4\sqrt{|x|}}\chi_{[-1,1]}(x).$$

Then, a simple calculation gives

$$P_J([a,b]) = \begin{cases} b - a - \dfrac{2}{3}(b-a)^{\frac{3}{2}} & \text{if } b - a < 1, \\[2mm] \dfrac{1}{3} & \text{if } b - a \geq 1. \end{cases} \tag{1.25}$$

Now $C_J = 6$ and, on account of (1.22) and (1.25), for ϵ small, we have

$$\frac{C_J}{\epsilon}P_{J_\epsilon}([a,b]) = \frac{6}{\epsilon}P_{J_\epsilon}([a,b]) = 6P_J\left(\left[\frac{a}{\epsilon},\frac{b}{\epsilon}\right]\right) = 2 = \mathrm{Per}([a,b]).$$

Therefore, in these two examples we have obtained that

$$\lim_{\epsilon\downarrow 0}C_{J_\epsilon}P_{J_\epsilon}(E) = \mathrm{Per}(E).$$

The following results by J. Dávila and A. Ponce will be especially important in order to get the rescaling results. See also [17, 20, 62, 72].

Theorem 1.8 (Dávila [36]) *Let $B \subset \mathbb{R}^N$ be open, bounded with a Lipschitz boundary, and let $0 \leq \rho_\epsilon$ be the radial functions satisfying*

$$\int_{\mathbb{R}^N}\rho_\epsilon(x)dx = 1, \qquad \lim_{\epsilon\to 0}\int_{|x|>\delta}\rho_\epsilon(x)dx = 0 \quad \forall\delta > 0. \tag{1.26}$$

Then

$$\lim_{\epsilon\to 0}\int_B\int_B\frac{|u(x)-u(y)|}{|x-y|}\rho_\epsilon(x-y)dxdy = K_{1,N}\int_B|Du|,$$

where

$$K_{1,N} = \frac{1}{N\omega_N}\int_{S^{N-1}}|e\cdot\sigma|d\sigma = \frac{\Gamma\left(\frac{N}{2}\right)}{\sqrt{\pi}\,\Gamma\left(\frac{N+1}{2}\right)}, \qquad |e| = 1. \tag{1.27}$$

Theorem 1.9 (Ponce [63]) *Let $N \geq 2$. Let $B \subset \mathbb{R}^N$ be open, bounded with a Lipschitz boundary, and let $0 \leq \rho_\epsilon$ be the radial functions satisfying (1.26). Let $\epsilon_n \downarrow 0$ as $n \to +\infty$. If $\{u_n\}_n \subset L^1(\Omega)$ is a bounded sequence such that*

$$\int_B \int_B \frac{|u_n(x) - u_n(y)|}{|x - y|} \rho_{\epsilon_n}(x - y) dx dy \leq K_{1,N} M,$$

where $K_{1,N}$ is given in (1.27) and M is a constant, then $\{u_n\}_n$ is relatively compact in $L^1(B)$. Moreover, if $u_{n_j} \to u$ in $L^1(B)$, then $u \in BV(B)$ and

$$\int_B |Du| \leq M.$$

The previous result holds in dimension $N \geq 2$; a counterexample for dimension $N = 1$ appears in [17].

Remark 1.10 For J satisfying (1.19),

$$\frac{M_J}{2} C_J = \frac{1}{K_{1,N}}. \tag{1.28}$$

Indeed, denoting $\tilde{J}(r) = J(x)$ if $|x| = r$,

$$K_{1,N} \frac{M_J}{2} C_J = K_{1,N} \frac{\int_{\mathbb{R}^N} |w| J(w) dw}{\int_{\mathbb{R}^N} |w_N| J(w) dw}$$

$$= K_{1,N} \frac{\int_0^\infty \int_{\partial B_r} r \tilde{J}(r) d\sigma dr}{\int_0^\infty \int_{\partial B_r} \tilde{J}(r) |\sigma \cdot e_N| d\sigma dr}$$

$$= \frac{\int_{S^{N-1}} |e_N \cdot \sigma| d\sigma}{N \omega_N} \cdot \frac{N \omega_N \int_0^\infty r^N \tilde{J}(r) dr}{\int_0^\infty \tilde{J}(r) r^N \int_{S^{N-1}} |e_N \cdot \sigma| d\sigma dr} = 1.$$

Theorem 1.11 *Assume J satisfies condition (1.19). If $u \in BV(\mathbb{R}^N)$ has compact support, then*

$$\lim_{\epsilon \downarrow 0} C_{J_\epsilon} \mathcal{TV}_{J_\epsilon}(u) = \int_{\mathbb{R}^N} |Du|.$$

In particular, if $E \subset \mathbb{R}^N$ is a bounded set of finite perimeter, then

$$\lim_{\epsilon \downarrow 0} C_{J_\epsilon} P_{J_\epsilon}(E) = \lim_{\epsilon \downarrow 0} \frac{C_J}{\epsilon} \epsilon^N P_J \left(\frac{E}{\epsilon} \right) = \text{Per}(E).$$

Proof Since u has compact support, for a large ball B containing supp(u) we can rewrite

$$TV_{J_\epsilon}(u) = \frac{1}{2} \int_B \int_B J_\epsilon(x - y)|u(y) - u(x)|dydx,$$

and

$$\int_{\mathbb{R}^N} |Du| = \int_B |Du|.$$

Now,

$$C_{J_\epsilon} TV_{J_\epsilon}(u) = \frac{1}{K_{1,N}} \int_B \int_B \frac{|u(x) - u(y)|}{|x - y|} \rho_\epsilon(x - y)dxdy,$$

being

$$\rho_\epsilon(z) = \frac{1}{2} C_J K_{1,N} \frac{|z|}{\epsilon} J_\epsilon(z).$$

Then, by (1.28),

$$\int_{\mathbb{R}^N} \rho_\epsilon(z)dz = \frac{1}{2} C_J K_{1,N} \int_{\mathbb{R}^N} \frac{|z|}{\epsilon} \frac{1}{\epsilon^N} J \left(\frac{z}{\epsilon} \right) dz$$

$$= \frac{1}{2} C_J K_{1,N} \int_{\mathbb{R}^N} |z| J(z)dz$$

$$= \frac{1}{2} C_J K_{1,N} M_J$$

$$= 1,$$

and, for $\delta > 0$,

$$\lim_{\epsilon \to 0} \int_{|z|>\delta} \rho_\epsilon(z)dz = \frac{1}{2} C_J K_{1,N} \lim_{\epsilon \to 0} \int_{|z|>\frac{\delta}{\epsilon}} |z| J(z)dz = 0,$$

since J satisfies (1.19).

  Therefore, we can apply Theorem 1.8 to get the result.   □

Nonlocal Isoperimetric Inequality

2

2.1 Nonlocal Isoperimetric Inequality

For the nonlocal perimeter, there is also an isoperimetric inequality, and here the main hypothesis used on J is that it is radially nonincreasing.

Its proof uses the symmetric decreasing rearrangement, which replaces a given nonnegative function f by a radial function f^*. Let us recall briefly the definition and some basic properties of this rearrangement.

Let E be a measurable set of finite measure. Its symmetric rearrangement E^* is given by the open centred ball whose measure agrees with $|E|$, that is, $E^* = B_r$ if r is such that

$$|B_r| = \omega_N r^N = |E|.$$

Now for a nonnegative and measurable function f that vanishes at infinity, in the sense that all its positive level sets have finite measure, we define the symmetric decreasing rearrangement f^* by symmetrizing its level sets:

$$f^*(x) = \int_0^\infty \chi_{\{f(x)>t\}^*}\, dt.$$

Note that for a radially nonincreasing function, it holds that $f^* = f$ and that the previous definitions are consistent in the sense that

$$\chi_{A^*} = (\chi_A)^*.$$

We refer to [54, 56] for details.

© Springer Nature Switzerland AG 2019
J. M. Mazón et al., *Nonlocal Perimeter, Curvature and Minimal Surfaces for Measurable Sets*, Frontiers in Mathematics,
https://doi.org/10.1007/978-3-030-06243-9_2

For the rearrangement of functions interacting with a convolution, we have the Riesz rearrangement inequality [56, Theorem 3.7], namely,

$$\int_{\mathbb{R}^N} f(x)(g * h)(x)\, dx \le \int_{\mathbb{R}^N} f^*(x)(g^* * h^*)(x)\, dx. \tag{2.1}$$

Theorem 2.1 (Isoperimetric Inequality) *Assume J is radially nonincreasing. For every measurable set E with finite measure, it holds that*

$$P_J(E) \ge P_J(B_r) \tag{2.2}$$

where B_r is a ball such that $|B_r| = |E|$.

Proof Using (2.1), one has

$$P_J(E) = \int_E \left(\int_{\mathbb{R}^N \setminus E} J(x - y) dy \right) dx$$

$$= |E| - \int_E \left(\int_E J(x - y) dy \right) dx$$

$$= |E| - \int_{\mathbb{R}^N} \chi_E(x)(J * \chi_E)(x)\, dx$$

$$\ge |E| - \int_{\mathbb{R}^N} (\chi_E)^*(x)(J^* * (\chi_E)^*)(x)\, dx$$

$$= |B_r| - \int_{\mathbb{R}^N} \chi_{B_r}(x)(J * \chi_{B_r})(x)\, dx$$

$$= P^J(B_r),$$

and we conclude (2.2) from the fact that $|E| = |B_r|$.   □

We are going to get now the nonlocal version of the relative isoperimetric inequality (1.3) obtained by De Giorgi. We will use the following Poincaré-type inequality given in [11, Proposition 4.1].

Proposition 2.2 *Assume that J is radially nonincreasing. Given $q \ge 1$ and Ω a bounded domain in $\mathbb{R}^N$, the quantity:*

$$\beta_{q-1}(J, \Omega, q) = \inf_{u \in L^q(\Omega),\, \int_\Omega u = 0} \frac{\dfrac{1}{2} \displaystyle\int_\Omega \int_\Omega J(x - y)|u(y) - u(x)|^q\, dy\, dx}{\displaystyle\int_\Omega |u(x)|^q\, dx}$$

is strictly positive. Consequently,

$$\beta_{q-1}(J,\Omega,q)\int_{\Omega}\left|u-\frac{1}{|\Omega|}\int_{\Omega}u\right|^{q}\le\frac{1}{2}\int_{\Omega}\int_{\Omega}J(x-y)|u(y)-u(x)|^{q}\,dy\,dx, \qquad (2.3)$$

for every $u \in L^{q}(\Omega)$.

Proof It is enough to prove that there exists a constant c such that

$$\|u\|_{q}\le c\left(\left(\int_{\Omega}\int_{\Omega}J(x-y)|u(y)-u(x)|^{q}dydx\right)^{1/q}+\left|\int_{\Omega}u\right|\right), \qquad (2.4)$$

for every $u \in L^{q}(\Omega)$.

Since J is radially nonincreasing, there exists $r > 0$ such that

$$J(z) \ge \alpha > 0 \quad \text{in } B_{r} \setminus \{0\}.$$

Since $\overline{\Omega} \subset \cup_{x\in\Omega}B_{r/2}(x)$, there exists $\{x_i\}_{i=1}^{m} \subset \Omega$ such that $\Omega \subset \cup_{i=1}^{m}B_{r/2}(x_i)$. Let $0 < \delta < r/2$ such that $B_{\delta}(x_i) \subset \Omega$ for all $i = 1, \dots, m$. Then, for any $\hat{x}_i \in B_{\delta}(x_i)$, $i = 1, \dots, m$,

$$\Omega = \bigcup_{i=1}^{m}(B_{r}(\hat{x}_i) \cap \Omega). \qquad (2.5)$$

Let us argue by contradiction. Suppose that (2.4) is false. Then, there exists $u_n \in L^{q}(\Omega)$, with $\|u_n\|_{L^{q}(\Omega)} = 1$, and satisfying

$$1\ge n\left(\left(\int_{\Omega}\int_{\Omega}J(x-y)|u_n(y)-u_n(x)|^{q}dydx\right)^{1/q}+\left|\int_{\Omega}u_n\right|\right) \quad \forall n \in \mathbb{N}.$$

Consequently,

$$\lim_{n}\int_{\Omega}\int_{\Omega}J(x-y)|u_n(y)-u_n(x)|^{q}\,dy\,dx = 0 \qquad (2.6)$$

and

$$\lim_{n}\int_{\Omega}u_n = 0. \qquad (2.7)$$

Let

$$F_n(x, y) = J(x - y)^{1/q} |u_n(y) - u_n(x)|$$

and

$$f_n(x) = \int_\Omega J(x - y)|u_n(y) - u_n(x)|^q \, dy.$$

From (2.7), it follows that

$$f_n \to 0 \quad \text{in } L^1(\Omega).$$

Passing to a subsequence if necessary, we can assume that

$$f_n(x) \to 0 \quad \forall x \in \Omega \setminus D_1, \quad D_1 \text{ null.} \tag{2.8}$$

On the other hand, by (2.6), we also have that

$$F_n \to 0 \quad \text{en } L^q(\Omega \times \Omega).$$

So, we can suppose, up to a subsequence,

$$F_n(x, y) \to 0 \quad \forall (x, y) \in \Omega \times \Omega \setminus C, \quad C \text{ null.} \tag{2.9}$$

Let $D_2 \subset \Omega$ be a null set satisfying that,

$$\text{for all } x \in \Omega \setminus D_2, \text{ the section } C_x \text{ of } C \text{ is null.} \tag{2.10}$$

Let $\hat{x}_1 \in B_\delta(x_1) \setminus (D_1 \cup D_2)$, then there exists a subsequence, denoted equal, such that

$$u_n(\hat{x}_1) \to \lambda_1 \in [-\infty, +\infty].$$

Consider now $\hat{x}_2 \in B_\delta(x_2) \setminus (D_1 \cup D_2)$, then up to a subsequence, we can assume

$$u_n(\hat{x}_2) \to \lambda_2 \in [-\infty, +\infty].$$

So, successively, for $\hat{x}_m \in B_\delta(x_m) \setminus (D_1 \cup D_2)$, there exists a subsequence, again denoted equal, such that

$$u_n(\hat{x}_m) \to \lambda_m \in [-\infty, +\infty].$$

By (2.9) and (2.10),

$$u_n(y) \to \lambda_i \quad \forall y \in (B_r(\hat{x}_i) \cap \Omega) \setminus C_{\hat{x}_i}.$$

Now, by (2.5),

$$\Omega = (B_r(\hat{x}_1) \cap \Omega) \cup (\cup_{i=2}^m (B_r(\hat{x}_i) \cap \Omega)).$$

Hence, since Ω is a bounded domain, there exists $i_2 \in \{2, .., m\}$ such that

$$(B_r(\hat{x}_1) \cap \Omega) \cap (B_r(\hat{x}_{i_2}) \cap \Omega) \neq \emptyset.$$

Therefore, $\lambda_1 = \lambda_{i_2}$. Let us call $i_1 = 1$. Again, since

$$\Omega = (B_r(\hat{x}_{i_1}) \cap \Omega) \cup (B_r(\hat{x}_{i_1}) \cap \Omega) \cup (\cup_{i \in \{1,\dots,m\} \setminus \{i_1, i_2\}} (B_r(\hat{x}_i) \cap \Omega)),$$

and there exists $i_3 \in \{1, \dots, m\} \setminus \{i_1, i_2\}$ such that

$$(B_r(\hat{x}_{i_1}) \cap \Omega) \cup (B_r(\hat{x}_{i_1}) \cap \Omega) \cap (B_r(\hat{x}_{i_3}) \cap \Omega) \neq \emptyset.$$

Consequently,

$$\lambda_{i_1} = \lambda_{i_2} = \lambda_{i_3}.$$

Using the same argument, we get

$$\lambda_1 = \lambda_2 = \dots = \lambda_m = \lambda.$$

If $|\lambda| = +\infty$, we have shown that

$$|u_n(y)|^q \to +\infty \quad \text{for almost every } y \in \Omega,$$

which contradicts $\|u_n\|_{L^q(\Omega)} = 1$ for all $n \in \mathbb{N}$. Hence, λ is finite.
On the other hand, by (2.8), $f_n(\hat{x}_i) \to 0$, $i = 1, \dots, m$. Hence,

$$F_n(\hat{x}_1, .) \to 0 \quad \text{in } L^q(\Omega).$$

Since $u_n(\hat{x}_1) \to \lambda$, from the above we conclude that

$$u_n \to \lambda \quad \text{in } L^q(B_r(\hat{x}_i) \cap \Omega).$$

Using again a compactness argument, we get

$$u_n \to \lambda \quad \text{in } L^q(\Omega).$$

By (2.7), $\lambda = 0$, so

$$u_n \to 0 \quad \text{in } L^q(\Omega),$$

which contradicts $\|u_n\|_{L^q(\Omega)} = 1$.   □

If we define

$$(u)_\Omega = \frac{1}{|\Omega|} \int_\Omega u(x) dx,$$

the Poincaré inequality (2.3), for the particular case $q = 1$ and $\Omega = B_\rho(x_0)$, can be written as:

$$\beta_0(J, B_\rho(x_0)) \int_{B_\rho(x_0)} |u - (u)_\Omega| \le \mathcal{TV}_J(u, B_\rho(x_0)), \tag{2.11}$$

for all $u \in L^1(B_\rho(x_0))$, where

$$\beta_0(J, B_\rho(x_0)) = \inf_{u \in L^1(B_\rho(x_0)), \int_{B_\rho(x_0)} u = 0} \frac{\mathcal{TV}_J(u, B_\rho(x_0))}{\int_{B_\rho(x_0)} |u(z)| \, dz}$$

We have that

$$\beta_0(J, B_\rho(x_0)) = \beta_0(J, B_\rho)$$

does not depend on x_0, and it is easy to see that

$$\beta_0(J, B_\rho) = \beta_0(J_{\frac{1}{\rho}}, B_1). \tag{2.12}$$

Following the proof of [6, Theorem 3.46], we give the following relative isoperimetric inequality.

Theorem 2.3 (Relative Isoperimetric Inequality) *Assume that J is radially nonincreasing and satisfies condition* (1.19). *Let $N \ge 2$ be an integer. For any set E of finite J-perimeter in $\mathbb{R}^N$, either E or $\mathbb{R}^N \setminus E$ has finite Lebesgue measure and*

$$P_J(B_{\rho_E}) \le P_J(E) \quad \text{for } |B_{\rho_E}| = \min\left\{ |E|, |\mathbb{R}^N \setminus E| \right\}.$$

Proof Applying (2.11) to $u = \chi_E$, and having in mind (2.12), we get

$$2(\chi_E)_{B_\rho(x)}(1 - (\chi_E)_{B_\rho(x)}) \leq \frac{1}{\omega_N \rho^N \beta_0(J_{1/\rho}, B_1)} TV_J(\chi_E, B_\rho(x)).$$

Since $(\chi_E)_{B_\rho(x)} \in [0, 1]$ and $\min\{t, 1 - t\} \leq 2t(1 - t)$ for any $t \in [0, 1]$, we obtain

$$\min\left\{(\chi_E)_{B_\rho(x)}, 1 - (\chi_E)_{B_\rho(x)}\right\} \leq \frac{1}{\omega_N \rho^N \beta_0(J_{1/\rho}, B_1)} TV_J(\chi_E, B_\rho(x)), \tag{2.13}$$

which implies

$$\min\left\{(\chi_E)_{B_\rho(x)}, 1 - (\chi_E)_{B_\rho(x)}\right\} \leq \frac{1}{\omega_N \rho^N \beta_0(J_{1/\rho}, B_1)} P_J(E). \tag{2.14}$$

Let us see now that

$$\{\rho^N \beta_0(J_{1/\rho}, B_1) \; : \; \rho > 0\} \quad \text{is not bounded.} \tag{2.15}$$

In fact, if there exists $M > 0$ such that $\rho^N \beta_0(J_{1/\rho}, B_1) \leq M$ for all $\rho > 0$, we have

$$\inf_{\int_{B_1} u = 0, \; \int_{B_1} |u| = 1} \frac{1}{2} \int_{B_1} \int_{B_1} J_{\frac{1}{\rho}}(x - y) \frac{|u(y) - u(x)|}{1/\rho} dy dx \leq M(1/\rho)^{N-1}.$$

Then, we can find u_ρ such that

$$\int_{B_1} u_\rho = 0, \quad \int_{B_1} |u_\rho| = 1, \tag{2.16}$$

and

$$\frac{1}{2} \int_{B_1} \int_{B_1} J_{\frac{1}{\rho}}(x - y) \frac{|u_\rho(y) - u_\rho(x)|}{1/\rho} dy dx \leq 2M(1/\rho)^{N-1}. \tag{2.17}$$

From (2.17), by Theorem 1.9, we get $\rho_n \to +\infty$ such that

$$u_{\rho_n} \to u \quad \text{in } L^1(B_1),$$

$$u \in BV(B_1),$$

$$\int_{B_1} |Du| = 0.$$

On the other hand, by (2.16), we have

$$\int_{B_1} u = 0 \quad \text{and} \quad \int_{B_1} |u| = 1.$$

and we arrive to a contradiction. Therefore, (2.15) holds.

Then, there exists $\rho_0 > 0$, depending on $P_J(E)$, such that

$$\frac{1}{\omega_N \rho_0{}^N \beta_0(J_{1/\rho_0}, B_1)} P_J(E) < \frac{1}{2}.$$

Hence by (2.14),

$$(\chi_E)_{B_{\rho_0}(x)} \in (0, 1/2) \cup (1/2, 1).$$

By a continuity argument, either

$$(\chi_E)_{B_{\rho_0}(x)} \in (0, 1/2) \text{ for any } x \in \mathbb{R}^N$$

or

$$(\chi_E)_{B_{\rho_0}(x)} \in (1/2, 1) \text{ for any } x \in \mathbb{R}^N.$$

If the first possibility is true, by (2.13), we obtain

$$\frac{|E \cap B_{\rho_0}(x)|}{\omega_N \rho_0^N} = (\chi_E)_{B_{\rho_0}(x)} \le \frac{1}{\omega_N \rho_0^N \beta_0(J_{1/\rho_0}, B_1)} TV_J(\chi_E, B_{\rho_0}(x)).$$

Hence,

$$|E \cap B_{\rho_0}(x)| \le \frac{TV_J(\chi_E, B_{\rho_0}(x))}{\beta_0(J_{1/\rho_0}, B_1)}. \tag{2.18}$$

For $i = 1, 2, \ldots, k$, let F_i be a numerable family of disjoint balls of radius ρ_0, such that the union of the members of this family covers $\mathcal{L}^N$-almost all of $\mathbb{R}^N$. Then, by (2.18), we have

$$|E| \le \sum_{i=1}^k \sum_{B \in F_i} |E \cap B|$$

$$\le \frac{1}{\beta_0(J_{1/\rho_0}, B_1)} \sum_{i=1}^k \sum_{B \in F_i} TV_J(E, B)$$

$$\le \frac{1}{\beta_0(J_{1/\rho_0}, B_1)} \sum_{i=1}^k TV_J(E, \mathbb{R}^N),$$

and so

$$|E| \le \frac{k}{\beta_0(J_{1/\rho_0}, B_1)} P_J(E).$$

If $(\chi_E)_{B_{\rho_0}(x)} \in (1/2, 1)$ for any $x \in \mathbb{R}^N$, a symmetric argument shows that $|\mathbb{R}^N \setminus E|$ can be estimated as above.

This shows that either E or $\mathbb{R}^N \setminus E$ has finite Lebesgue measure. Finally, the second part of the result is consequence of the Isoperimetric Inequality given in Theorem 2.1. $\square$

In Sect. 3.3, we will give an isoperimetric inequality involving the nonlocal curvature.

2.2   Nonlocal Coarea Formula

Similarly to the coarea formula (1.8) in the local case, we have the following nonlocal coarea formula.

Theorem 2.4 (Coarea Formula)   *For any $u \in L^1(\mathbb{R}^N)$, let*

$$E_t(u) = \{x \in \mathbb{R}^N \ : \ u(x) > t\}.$$

Then,

$$\mathcal{TV}_J(u) = \int_{-\infty}^{+\infty} P_J(E_t(u)) \, dt. \tag{2.19}$$

Proof Since

$$u(x) = \int_0^\infty \chi_{E_t(u)}(x) \, dt - \int_{-\infty}^0 (1 - \chi_{E_t(u)}(x)) \, dt,$$

we have

$$u(y) - u(x) = \int_{-\infty}^{+\infty} \chi_{E_t(u)}(y) - \chi_{E_t(u)}(x) \, dt.$$

Moreover, since $u(y) \ge u(x)$ implies $\chi_{E_t(u)}(y) \ge \chi_{E_t(u)}(x)$, we obtain that

$$|u(y) - u(x)| = \int_{-\infty}^{+\infty} |\chi_{E_t(u)}(y) - \chi_{E_t(u)}(x)| \, dt.$$

Therefore, by Tonelli–Hobson's Theorem, we get

$$\mathcal{TV}_J(u) = \frac{1}{2} \int_{\mathbb{R}^N} \int_{\mathbb{R}^N} J(x - y) |u(y) - u(x)| dx dy$$

$$= \frac{1}{2} \int_{\mathbb{R}^N} \int_{\mathbb{R}^N} J(x - y) \left(\int_{-\infty}^{+\infty} |\chi_{E_t(u)}(y) - \chi_{E_t(u)}(x)| dt \right) dx dy$$

$$= \int_{-\infty}^{+\infty} \left(\frac{1}{2} \int_{\mathbb{R}^N} \int_{\mathbb{R}^N} J(x - y) |\chi_{E_t(u)}(y) - \chi_{E_t(u)}(x)| dx dy \right) dt$$

$$= \int_{-\infty}^{+\infty} P_J(E_t(u)) dt.$$

$\square$

Nonlocal Minimal Surfaces and Nonlocal Curvature

3

Recall that if a set E has minimal local perimeter in a bounded set Ω, then it has zero mean curvature at each point of $\partial E \cap \Omega$ (see [51]), and the equation that says that the curvature is equal to zero is the Euler–Lagrange equation associated to the minimization of the perimeter of a set.

In order to give a nonlocal version of this result we will also introduce the nonlocal concept of mean curvature.

3.1 Nonlocal Minimal Surfaces and Nonlocal Curvature

Let Ω be an open bounded subset of $\mathbb{R}^N$. We say that a measurable set $E \subset \mathbb{R}^N$ is *J-minimal* in Ω if

$$P_J(E, \Omega) \leq P_J(F, \Omega) \text{ for any measurable set } F \text{ such that } F \setminus \Omega = E \setminus \Omega.$$

Proposition 3.1 *Let Ω be an open bounded subset of $\mathbb{R}^N$. Then, $E \subset \mathbb{R}^N$ is J-minimal in Ω if and only if*

$$P_J(E) \leq P_J(F) \text{ for any measurable set } F \text{ such that } F \setminus \Omega = E \setminus \Omega.$$

Proof If $E \setminus \Omega = F \setminus \Omega$, then $\mathbb{R}^N \setminus (\Omega \cup E) = \mathbb{R}^N \setminus (\Omega \cup F)$. Therefore, since

$$P_J(A, \Omega) = L_J(A, \mathbb{R}^N \setminus A) - L_J(A \setminus \Omega, \mathbb{R}^N \setminus (A \cup \Omega))$$

we get the result. $\qquad\square$

© Springer Nature Switzerland AG 2019
J. M. Mazón et al., *Nonlocal Perimeter, Curvature and Minimal Surfaces for Measurable Sets*, Frontiers in Mathematics,
https://doi.org/10.1007/978-3-030-06243-9_3

We now introduce the concept on nonlocal curvature.

Definition 3.2 Let $E \subset \mathbb{R}^N$ be measurable, with $N \geq 2$. For a point $x \in \mathbb{R}^N$ we define the *J-mean curvature* of ∂E at x as

$$H^J_{\partial E}(x) = -\int_{\mathbb{R}^N} J(x - y)(\chi_E(y) - \chi_{\mathbb{R}^N \setminus E}(y))dy.$$

Some observations are in order. First of all, observe that

$$H^J_{\partial E}(x) = -H^J_{\partial(\mathbb{R}^N \setminus E)}(x),$$

and that

$$-1 \leq H^J_{\partial E}(x) \leq 1.$$

Note also that $H^J_{\partial E}(x)$ makes perfect sense for every $x \in \mathbb{R}^N$, not necessary for points in ∂E. This fact will be used later in the book.

Like the usual mean curvature, if ∂E is a smooth boundary, for $x \in \partial E$, $H^J_{\partial E}(x)$ measures in some average sense the deviation of ∂E from its tangent hyperplane at x.

Let us point out finally that with our signs choice, the curvature of a ball is positive as it is commonly in geometry texts. Some authors define the curvature with the reverse sign, see, for example, [1, 29].

Definition 3.3 We say that ∂E is a *J-minimal surface* in a bounded open set Ω if the set $\partial E \cap \Omega$ satisfies the nonlocal minimal surface equation

$$H^J_{\partial E}(x) = 0$$

for all $x \in \partial E \cap \Omega$ such that

$$|E \cap B_\delta(x)| > 0 \text{ and } |(\mathbb{R}^N \setminus E) \cap B_\delta(x)| > 0 \text{ for every small } \delta > 0.$$

As examples of nonlocal minimal surfaces we have the following. By symmetry, any hyperplane

$$\{x \in \mathbb{R}^N : x_N > 0\}$$

is a J-minimal surface in $\mathbb{R}^N$. Also, the classical cone in the plane,

$$\{(x, y) \in \mathbb{R}^2 : xy > 0\},$$

is a J-minimal surface in $\mathbb{R}^2$.

Theorem 3.4 *Let Ω be an open bounded subset of $\mathbb{R}^N$. If $E \subset \mathbb{R}^N$ is a J-minimal set in Ω, then ∂E is a J-minimal surface in Ω.*

Proof Let $E \subset \mathbb{R}^N$ be a J-minimal set in Ω. Given $x_0 \in \partial E \cap \Omega$, there is $\delta > 0$ such that $B_\delta(x_0) \subset \Omega$. Now consider

$$A_\delta = B_\delta(x_0) \cap (\mathbb{R}^N \setminus E)$$

and

$$F_\delta = A_\delta \cup E.$$

Since, $F_\delta \setminus \Omega = E \setminus \Omega$, we have

$$P_J(E) \le P_J(F_\delta),$$

that is,

$$\int_E \int_{\mathbb{R}^N \setminus E} J(x-y) dy dx \le \int_{F_\delta} \int_{\mathbb{R}^N \setminus F_\delta} J(x-y) dy dx.$$

Now, since $A_\delta \cap E = \emptyset$,

$$\int_E \int_{\mathbb{R}^N \setminus E} J(x-y) dy dx = \int_{F_\delta} \int_{\mathbb{R}^N \setminus E} J(x-y) dy dx - \int_{A_\delta} \int_{\mathbb{R}^N \setminus E} J(x-y) dy dx.$$

Therefore

$$\int_{A_\delta} \int_{\mathbb{R}^N \setminus E} J(x-y) dy dx \ge \int_{F_\delta} \left(\int_{\mathbb{R}^N \setminus E} J(x-y) dy - \int_{\mathbb{R}^N \setminus F_\delta} J(x-y) dy \right) dx$$

$$= \int_{F_\delta} \left(\int_{A_\delta} J(x-y) dy \right) dx$$

$$= \int_{A_\delta} \int_{A_\delta} J(x-y) dy dx + \int_E \int_{A_\delta} J(x-y) dy dx.$$

Hence

$$\int_{A_\delta} \int_{A_\delta} J(x-y) dy dx \le \int_{A_\delta} \left(\int_{\mathbb{R}^N \setminus E} J(x-y) dy - \int_E J(x-y) dy \right) dx.$$

Now, since $x_0 \in \partial E$, we have $|A_\delta| > 0$. Then, dividing by $|A_\delta|$, using that

$$x \mapsto \left(\int_{\mathbb{R}^N \setminus E} J(x - y) dy - \int_E J(x - y) dy \right)$$

is continuous, and letting $\delta \to 0$, we conclude that

$$\left(\int_{\mathbb{R}^N \setminus E} J(x_0 - y) dy - \int_E J(x_0 - y) dy \right) \geq 0.$$

With a similar procedure, but taking now $\tilde{A}_\delta = B_\delta(x_0) \cap E$ and $\tilde{F}_\delta = E \setminus \tilde{A}_\delta$, we arrive to

$$\int_{\tilde{A}_\delta} \left(\int_{\mathbb{R}^N \setminus E} J(x - y) dy - \int_E J(x - y) dy \right) dx \leq - \int_{\tilde{A}_\delta} \int_{\tilde{A}_\delta} J(x - y) dy dx.$$

Then, dividing by $|\tilde{A}_\delta| > 0$ and letting $\delta \to 0$, we get

$$\left(\int_{\mathbb{R}^N \setminus E} J(x_0 - y) dy - \int_E J(x_0 - y) dy \right) \leq 0.$$

Therefore, as we wanted to show,

$$H_{\partial E}^J(x_0) = - \int_{\mathbb{R}^N} J(x_0 - y)(\chi_E(y) - \chi_{\mathbb{R}^N \setminus E}(y)) dy = 0.$$

$\square$

Remark 3.5 Let us point out that a similar result to Theorem 3.4 has been obtained in [1, 29] for the singular kernel $\frac{1}{|x|^{N+s}}$, where the nonlocal curvature at $x \in \partial E$, for E with smooth boundary, is defined (formally) as

$$H_{\partial E}^s(x) = - \int_{\mathbb{R}^N} \frac{\chi_E(y) - \chi_{\mathbb{R}^N \setminus E}(y)}{|x - y|^{N+s}} dy.$$

Moreover in this case the following rescaling formula holds for smooth sets:

$$\lim_{s \to 1} (1 - s) H_{\partial E}^s(x) = (N - 1) \omega_{N-1} H_{\partial E}(x),$$

where $H_{\partial E}(x)$ is the (local) mean curvature of ∂E at x, see [1].

3.2 Rescaling the Nonlocal Curvature

In this section we will give a convergence result of the nonlocal J-curvature under a rescale of the kernel. First of all we present two examples for illustration.

Example 3.6

1. If $J(x) = \frac{1}{|B_1|} \chi_{B_1}$, then for any $x \in \overline{B_r(x_0)}$, we have

$$H^J_{\partial B_r(x_0)}(x) = -\int_{\mathbb{R}^N} J(x - y)(\chi_{B_r(x_0)}(y) - \chi_{\mathbb{R}^N \setminus B_r(x_0)}(y))dy$$

$$= -\frac{1}{|B_1|}\int_{B_1(x)}(\chi_{B_r(x_0)}(y) - \chi_{\mathbb{R}^N \setminus B_r(x_0)}(y))dy$$

$$= -\frac{1}{|B_1|}\left(|B_1(x) \cap B_r(x_0)| - |B_1(x) \cap (\mathbb{R}^N \setminus B_r(x_0))|\right)$$

$$= -\frac{1}{|B_1|}(2|B_1(x) \cap B_r(x_0)| - |B_1(x)|)$$

$$= -\frac{2|B_1(x) \cap B_r(x_0)|}{|B_1|} + 1.$$

And for the rescaled kernel,

$$H^{J_\epsilon}_{\partial B_r(x_0)}(x) = -\frac{1}{\epsilon^N}\int_{\mathbb{R}^N} J\left(\frac{x-y}{\epsilon}\right)(\chi_{B_r(x_0)}(y) - \chi_{\mathbb{R}^N \setminus B_r(x_0)}(y))dy$$

$$= -\frac{2|B_\epsilon(x) \cap B_r(x_0)|}{|B_\epsilon|} + 1.$$

Now, for $N = 2$, a simple calculus gives, for $x \in \partial B_r(x_0)$ and ϵ small,

$$|B_\epsilon(x) \cap B_r(x_0)| = \epsilon^2\left[-\frac{\epsilon}{2r}\sqrt{1 - \left(\frac{\epsilon}{2r}\right)^2} + \arcsin\left(-\frac{\epsilon}{2r}\right) + \frac{\pi}{2}\right]$$

$$+ r^2\left[\frac{\pi}{2} - \frac{2r^2 - \epsilon^2}{2r^2}\sqrt{1 - \left(\frac{2r^2 - \epsilon^2}{2r^2}\right)^2} - \arcsin\left(\frac{2r^2 - \epsilon^2}{2r^2}\right)\right].$$

Hence

$$H_{\partial B_r(x_0)}^{J_\epsilon}(x) = \frac{\epsilon}{\pi r}\sqrt{1 - \left(\frac{\epsilon}{2r}\right)^2} + \frac{2}{\pi}\arcsin\left(\frac{\epsilon}{2r}\right)$$

$$-\frac{2r^2}{\pi\epsilon^2}\left[\frac{\pi}{2} - \frac{2r^2 - \epsilon^2}{2r^2}\sqrt{1 - \left(\frac{2r^2 - \epsilon^2}{2r^2}\right)^2} - \arcsin\left(\frac{2r^2 - \epsilon^2}{2r^2}\right)\right].$$

Now,

$$C_J = \frac{2}{\displaystyle\int_{\mathbb{R}^2} J(z)|z_2|dz} = \frac{2}{\displaystyle\frac{1}{|B_1|}\int_{B_1}|z_2|dz} = \frac{3\pi}{2}.$$

Therefore,

$$\frac{C_J}{\epsilon}H_{\partial B_r(x_0)}^{J_\epsilon}(x) = \frac{3}{2r}\sqrt{1 - \left(\frac{\epsilon}{2r}\right)^2} + \frac{3}{\epsilon}\arcsin\left(\frac{\epsilon}{2r}\right)$$

$$-\frac{3r^2}{\epsilon^3}\left[\frac{\pi}{2} - \frac{2r^2 - \epsilon^2}{2r^2}\sqrt{1 - \left(\frac{2r^2 - \epsilon^2}{2r^2}\right)^2} - \arcsin\left(\frac{2r^2 - \epsilon^2}{2r^2}\right)\right].$$

Then, since

$$\lim_{\epsilon \to 0}\frac{3}{\epsilon}\arcsin\left(\frac{\epsilon}{2r}\right) = \frac{3}{2r}$$

and

$$\lim_{\epsilon \to 0}\frac{3r^2}{\epsilon^3}\left[\frac{\pi}{2} - \frac{2r^2 - \epsilon^2}{2r^2}\sqrt{1 - \left(\frac{2r^2 - \epsilon^2}{2r^2}\right)^2} - \arcsin\left(\frac{2r^2 - \epsilon^2}{2r^2}\right)\right] = \frac{2}{r},$$

we get

$$\lim_{\epsilon \to 0}\frac{C_J}{\epsilon}H_{\partial B_r(x_0)}^{J_\epsilon}(x) = \frac{3}{2r} + \frac{3}{2r} - \frac{2}{r} = \frac{1}{r} = H_{\partial B_r(x_0)}.$$

2. Consider now E as the square $E = \{(x, y) \in \mathbb{R}^2 : \|(x, y)\|_\infty \leq 1\}$ and the same kernel as above. Then, for $0 < \epsilon < 1$ we have

$$\frac{C_J}{\epsilon}H_{\partial E}^{J_\epsilon}(1, 1) = -\frac{3}{2\epsilon^3}\left(|E \cap B_\epsilon(1, 1)| - |(\mathbb{R}^2 \setminus E \cap B_\epsilon(1, 1)|\right) = \frac{3\pi}{4\epsilon}.$$

Therefore,

$$\lim_{\epsilon \to 0} \frac{C_J}{\epsilon} H_{\partial E}^{J_\epsilon}(1, 1) = +\infty.$$

Theorem 3.7 *Let $N \geq 2$. Assume that J has compact support $\overline{B_r}$, is continuous almost everywhere, and bounded. Let $E \subset \mathbb{R}^N$ be a smooth set such that ∂E is of class C^2. Then, for every $x \in \partial E$, we have*

$$\lim_{\epsilon \downarrow 0} C_{J_\epsilon} H_{\partial E}^{J_\epsilon}(x) = (N - 1) H_{\partial E}(x), \tag{3.1}$$

where $H_{\partial E}(x)$ is the (local) mean curvature of ∂E at x.

Proof It is well-known that curvature may be easily computed in normal coordinates. Namely, suppose ∂E is described as a graph in normal coordinates, meaning that, in an open ball B_{r_0}, ∂E coincides with the graph of a C^2 function $\varphi : B_{r_0} \cap \mathbb{R}^{N-1} \to \mathbb{R}$ with $\varphi(0) = 0$ and $\nabla \varphi(0) = 0$ such that $E \cap B_{r_0} = \{(y_1, \ldots, y_N) : y_N < \varphi(y_1, \ldots, y_{N-1})\}$. Since $D^2\varphi(0)$ is a real symmetric matrix, it will admit $N - 1$ real eigenvalues $\lambda_1, \ldots, \lambda_{N-1}$. Minus the arithmetic mean of the eigenvalues is called the mean curvature and we denote it by $H_{\partial E}(0)$, namely,

$$H_{\partial E}(0) = -\frac{\lambda_1 + \cdots + \lambda_{N-1}}{N - 1}.$$

We will assume that the coordinate axes $e_1, \ldots, e_{N-1}$ are the eigenvectors associated with the eigenvalues of $D^2\varphi(0)$, $\lambda_1, \ldots, \lambda_{N-1}$. We can also assume that $r_0 < r$. Then, for ϵ small enough,

$$C_{J_\epsilon} H_{\partial E}^{J_\epsilon}(0) = -C_{J_\epsilon} \int_{\mathbb{R}^N} J_\epsilon(y)(\chi_E(y) - \chi_{\mathbb{R}^N \setminus E}(y)) dy$$

$$= -\frac{C_{J_\epsilon}}{\epsilon^N} \int_{B_{r\epsilon}} J\left(\frac{y}{\epsilon}\right)(\chi_E(y) - \chi_{\mathbb{R}^N \setminus E}(y)) dy$$

$$= -\frac{C_{J_\epsilon}}{\epsilon^N} \left(\int_{\{y_N < \varphi(y_1, \ldots, y_{N-1})\} \cap B_{r\epsilon}} J\left(\frac{y}{\epsilon}\right) dy \right.$$

$$\left. - \int_{\{y_N > \varphi(y_1, \ldots, y_{N-1})\} \cap B_{r\epsilon}} J\left(\frac{y}{\epsilon}\right) dy \right).$$

Hence, changing variables as $z = \frac{y}{\epsilon}$, we get

$$C_{J_\epsilon} H_{\partial E}^{J_\epsilon}(0) = -C_{J_\epsilon}\left(\int_{\{z_N < \frac{1}{\epsilon}\varphi(\epsilon z_1,\ldots,\epsilon z_{N-1})\}\cap B_r} J(z)dz\right.$$

$$\left. - \int_{\{z_N > \frac{1}{\epsilon}\varphi(\epsilon z_1,\ldots,\epsilon z_{N-1})\}\cap B_r} J(z)dz\right).$$

And, since J is radially symmetric,

$$C_{J_\epsilon} H_{\partial E}^{J_\epsilon}(0) = -C_{J_\epsilon}\int_{\mathbb{R}^{N-1}\cap B_r}\int_{-\frac{1}{\epsilon}\varphi(\epsilon z_1,\ldots,\epsilon z_{N-1})}^{\frac{1}{\epsilon}\varphi(\epsilon z_1,\ldots,\epsilon z_{N-1})} J(z)dz.$$

Now, by Taylor's expansion, we have

$$\varphi(\epsilon z_1,\ldots,\epsilon z_{N-1}) = \frac{1}{2}D^2\varphi(0)(\epsilon z_1,\ldots,\epsilon z_{N-1}) + O(\epsilon^3)$$

$$= \frac{1}{2}\sum_{i=1}^{N-1}\lambda_i\epsilon^2 z_i^2 + O(\epsilon^3).$$

Therefore,

$$\lim_{\epsilon\to 0} C_{J_\epsilon} H_{\partial E}^{J_\epsilon}(0) = -\lim_{\epsilon\to 0} C_{J_\epsilon}\int_{\mathbb{R}^{N-1}\cap B_r}\int_{-\frac{1}{\epsilon}\varphi(\epsilon z_1,\ldots,\epsilon z_{N-1})}^{\frac{1}{\epsilon}\varphi(\epsilon z_1,\ldots,\epsilon z_{N-1})} J(z)dz.$$

$$\tag{3.2}$$

$$= -\lim_{\epsilon\to 0} C_J\int_{\mathbb{R}^{N-1}\cap B_r}\left[\frac{1}{\epsilon}\int_{-\left(\epsilon\frac{1}{2}\sum_{i=1}^{N-1}\lambda_i z_i^2 + O(\epsilon^2)\right)}^{\epsilon\frac{1}{2}\sum_{i=1}^{N-1}\lambda_i z_i^2 + O(\epsilon^2)} J(z)dz_N\right]dz_1 dz_2\ldots dz_{N-1}.$$

Then, since J is continuous almost everywhere and it is radially symmetric, we have

$$z_n \mapsto J(z_1, z_2, \ldots, z_{N-1}, z_N) \quad \text{is continuous at } 0 \tag{3.3}$$

for almost every $(z_1, z_2, \ldots, z_{N-1})$ in $\mathbb{R}^{N-1}$. Therefore

$$\lim_{\epsilon\to 0}\frac{1}{\epsilon}\int_{-\left(\epsilon\frac{1}{2}\sum_{i=1}^{N-1}\lambda_i z_i^2 + O(\epsilon^2)\right)}^{\epsilon\frac{1}{2}\sum_{i=1}^{N-1}\lambda_i z_i^2 + O(\epsilon^2)} J(z)dz_N = J(z_1, z_2, \ldots z_{N-1}, 0)$$

for almost every $(z_1, z_2, \ldots, z_{N-1})$ in $\mathbb{R}^{N-1}$. Now since J is bounded in B_r, we can pass to the limit in (3.2) to get

$$\lim_{\epsilon \to 0} C_{J_\epsilon} H_{\partial E}^{J_\epsilon}(0) = -C_J \sum_{i=1}^{N-1} \lambda_i \int_{\mathbb{R}^{N-1}} J(z_1, \ldots, z_{N-1}, 0) z_i^2 dz_1 \ldots dz_{N-1},$$

that is, on account of the symmetry of J,

$$\lim_{\epsilon \to 0} C_{J_\epsilon} H_{\partial E}^{J_\epsilon}(0) = -C_J \int_{\mathbb{R}^{N-1}} J(z_1, \ldots, z_{N-1}, 0) z_{N-1}^2 dz_1 \ldots dz_{N-1} \sum_{i=1}^{N-1} \lambda_i. \qquad (3.4)$$

Now, if we make the change of variables $z_i = z_i, i = 1, \ldots, N-2, z_{N-1} = r \cos\theta$, and $z_N = r \sin\theta$, we have, writing $\hat{z} = (z_1, z_2, \ldots, z_{N-2})$,

$$\int_{\mathbb{R}^N} J(z)|z_N| dz = \int_{\mathbb{R}^{N-2}} \int_0^{+\infty} \int_0^{2\pi} J(\hat{z}, r\cos\theta, r\sin\theta) r^2 |\sin\theta| d\theta dr d\hat{z}$$

$$= 4 \int_{\mathbb{R}^{N-2}} \int_0^{+\infty} J(\hat{z}, r, 0) r^2 dr d\hat{z}$$

$$= 2 \int_{\mathbb{R}^N} J(\hat{z}, r, 0) r^2 d\hat{z} dr$$

$$= 2 \int_{\mathbb{R}^{N-1}} J(z_1, \ldots, z_{N-1}, 0) z_{N-1}^2 dz_1 \ldots dz_{N-1}.$$

Hence

$$C_J \int_{\mathbb{R}^{N-1}} J(z_1, \ldots, z_{N-1}, 0) z_{N-1}^2 dz_1 \ldots dz_{N-1} = 1.$$

Therefore, from (3.4), we get

$$\lim_{\epsilon \to 0} C_{J_\epsilon} H_{\partial E}^{J_\epsilon}(0) = (N-1) H_{\partial E}(0).$$

$\square$

Remark 3.8 Observe, for example, that if J is radially non-increasing, then it is continuous almost everywhere. Let us explain why (3.3) is true. Take x as a point where J is continuous. Since J is radially symmetric, we have that J is also continuous at $(\hat{x}, 0) \in \mathbb{R}^N$ with $\hat{x} \in \mathbb{R}^{N-1}$ such that $|(\hat{x}, 0)| = |x|$. Now for

$$A = \{\hat{x} \in \mathbb{R}^{N-1} : J \text{ is continuous at } (\hat{x}, 0)\}$$

we have that $\mathcal{H}^{N-1}(\mathbb{R}^{N-1} \setminus A) = 0$ since, again, J is radially symmetric.

In the next example we will see that the assumption "∂E is of class C^2" in the above result is necessary.

Example 3.9 For $N = 2$. Assume that $J(x) = \frac{1}{|B_1|} \chi_{B_1}$. Let $\varphi : \mathbb{R} \to \mathbb{R}$ be the function defined by

$$\varphi(x) = \begin{cases} x^2 & \text{if } x \geq 0, \\ -x^2 & \text{if } x < 0, \end{cases}$$

and consider the set

$$E = \{(x, y) \in \mathbb{R}^2 \ : \ y < \varphi(x)\}.$$

Then by symmetry we have $H_{\partial E}^{J_\epsilon}(0, 0) = 0$, and consequently

$$\lim_{\epsilon \downarrow 0} C_{J_\epsilon} H_{\partial E}^{J_\epsilon}(0, 0) = 0.$$

Now, $H_{\partial E}(0, 0) = 2$ (here the curvature is understood as one over the largest radius of a tangent ball). Therefore, (3.1) is not true in this case.

3.3   Isoperimetric Inequalities and Curvature

Assume in this section that J is radially non-increasing.

Lemma 3.10 *The following relation holds true:*

$$\frac{d}{dr} P_J(B_r) = H_{\partial B_r}^J(y) \mathrm{Per}(B_r), \tag{3.5}$$

with $y \in \partial B_r$.

Proof It is enough to calculate the derivative from the right. So, for $h > 0$,

$$P_J(B_{r+h}) - P_J(B_r) = |B_{r+h}| - |B_r|$$

$$- \left(\iint_{B_{r+h} \times B_{r+h}} J(x - y) dy dx - \iint_{B_r \times B_r} J(x - y) dy dx \right).$$

Hence,

$$P_J(B_{r+h}) - P_J(B_r) = |B_{r+h}| - |B_r|$$

$$-\left(\iint_{(B_{r+h}\setminus B_r)\times(B_{r+h}\setminus B_r)} J(x-y)dydx - 2\iint_{(B_{r+h}\setminus B_r)\times B_r} J(x-y)dydx \right).$$

Now, we have

$$\lim_{h\to 0^+} \frac{|B_{r+h}| - |B_r|}{h} = \text{Per}(B_r),$$

and

$$\lim_{h\to 0^+} \frac{1}{h} \iint_{(B_{r+h}\setminus B_r)\times(B_{r+h}\setminus B_r)} J(x-y)dydx = 0.$$

Moreover

$$\lim_{h\to 0^+} \frac{1}{h} \iint_{(B_{r+h}\setminus B_r)\times B_r} J(x-y)dydx = \text{Per}(B_r)\frac{1 - H^J_{\partial B_r}(\tilde{y})}{2},$$

with $\tilde{y} \in \partial B_r$. In fact, making a spherical change of coordinates $y = g(\rho, \sigma)$ and having in mind the radial symmetry of the curvature, we have

$$\lim_{h\to 0^+} \frac{1}{h} \iint_{(B_{r+h}\setminus B_r)\times B_r} J(x-y)dydx$$

$$= \lim_{h\to 0^+} \frac{1}{h} \int_{B_r} \int_{S^{N-1}} \int_r^{r+h} \rho^{N-1} J(x - g(\rho,\sigma))d\rho d\sigma dx$$

$$= \int_{B_r} \int_{S^{N-1}} r^{N-1} J(x - g(r,\sigma))d\sigma dx$$

$$= r^{N-1} \int_{S^{N-1}} \int_{B_r} J(x - g(r,\sigma))dx d\sigma$$

$$= r^{N-1} \int_{S^{N-1}} \frac{1 - H^J_{\partial B_r}(g(r,\sigma))}{2} d\sigma$$

$$= r^{N-1} \int_{S^{N-1}} \frac{1 - H^J_{\partial B_r}(\tilde{y})}{2} d\sigma$$

$$= \text{Per}(B_r)\frac{1 - H^J_{\partial B_r}(\tilde{y})}{2},$$

for $\tilde{y} \in \partial B_r$.

Therefore, we get

$$\lim_{h \to 0^+} \frac{P_J(B_{r+h}) - P_J(B_r)}{h} = \text{Per}(B_r) H_{\partial B_r}^J(y),$$

with $y \in \partial B_r$.   □

Corollary 3.11

$$P_J(B_r) = \int_{B_r} H_{\partial B_{\|x\|}}^J(x) dx. \tag{3.6}$$

Proof By (3.5), we have

$$\frac{d}{dr} P_J(B_r) = H_{\partial B_r}^J(r, \overline{0}) \text{Per}(B_r). \tag{3.7}$$

Integrating in (3.7) we get

$$P_J(B_r) = N\omega_N \int_0^r \rho^{N-1} H_{\partial B_\rho}^J(\rho, \overline{0}) d\rho$$

$$= \int_{B_r} H_{\partial B_{\|x\|}}^J(x) dx.$$

□

Let E be a finite Lebesgue set. In the local case, the isoperimetric inequality says that

$$\text{Per}(B_{\rho_E}) \leq \text{Per}(E) \quad \text{for } |B_{\rho_E}| = |E|, \tag{3.8}$$

with equality holding if and only if E is a ball.
    Since

$$\rho_E = \left(\frac{|E|}{\omega_N}\right)^{\frac{1}{N}}$$

and

$$\text{Per}(B_{\rho_E}) = N\omega_N^{1/N} |E|^{\frac{N-1}{N}},$$

the above inequality can be written as

$$N\omega_N^{1/N} |E|^{\frac{N-1}{N}} \leq \text{Per}(E). \tag{3.9}$$

An isoperimetric type inequality for s-perimeters also holds (see, for example, [47]), and it states that

$$\text{Per}_s(B_{\rho_E}) \leq \text{Per}_s(E) \quad \text{for } |B_{\rho_E}| = |E|, \tag{3.10}$$

with equality holding if and only if E is a ball. Since

$$\text{Per}_s(B_r) = r^{N-s}\text{Per}_s(B_1), \tag{3.11}$$

the inequality (3.10) can be written as

$$\text{Per}_s(B_1)\left(\frac{|E|}{\omega_N}\right)^{\frac{N-s}{N}} \leq \text{Per}_s(E). \tag{3.12}$$

In our case, we can also obtain a similar inequality to (3.9) and (3.12). In fact, by (3.6)

$$P_J(B_r) = \int_{B_r} H^J_{\partial B_{\|x\|}}(x)dx.$$

Therefore, for $|B_{\rho_E}| = |E|$,

$$P_J(B_{\rho_E}) = N\omega_N \int_0^{(|E|/\omega_N)^{1/N}} r^{N-1} H^J_{\partial B_r}(r, \overline{0})dr,$$

and we can write the nonlocal isoperimetric inequality (2.2), assuming J is radially non-increasing, as

$$\psi_{J,N}(|E|) \leq P_J(E), \tag{3.13}$$

being $\psi_{J,N}$ the strictly increasing function

$$\psi_{J,N}(s) = N\omega_N \int_0^{(s/\omega_N)^{1/N}} r^{N-1} H^J_{\partial B_r}(r, \overline{0})dr. \tag{3.14}$$

But we can rewrite the above three isoperimetric inequalities with another common formulation. Assume J is radially non-increasing. First observe that, thanks to (3.6), we can write (2.2) as

$$\left(\frac{1}{|B_{\rho_E}|}\int_{B_{\rho_E}} H^J_{\partial B_{\|x\|}}(x)dx\right)|E| \leq P_J(E). \tag{3.15}$$

And this expression is the nonlocal version of the classical isoperimetric inequality (3.9), since it can be written as

$$\left(\frac{1}{|B_{\rho E}|} \int_{B_{\rho E}} \frac{N-1}{\|x\|} dx \right) |E| \leq \operatorname{Per}(E). \tag{3.16}$$

Observe that, if we consider the rescaled kernel J_ϵ, we have

$$\left(\frac{1}{|B_{\rho E}|} \int_{B_{\rho E}} H_{\partial B_{\|x\|}}^{J_\epsilon}(x) dx \right) |E| \leq P_{J_\epsilon}(E). \tag{3.17}$$

Now, assuming the assumptions of Theorems 1.11 and 3.7, we have

$$\lim_{\epsilon \downarrow 0} C_{J_\epsilon} P_{J_\epsilon}(E) = \operatorname{Per}(E)$$

and

$$\lim_{\epsilon \downarrow 0} C_{J_\epsilon} H_{\partial B_{\|x\|}}^{J_\epsilon}(x) = (N-1) H_{\partial B_{\|x\|}}(x) = \frac{N-1}{\|x\|}.$$

Therefore, taking limits in (3.17) as $\epsilon \to 0$ and applying the dominate convergence theorem, we obtain the isoperimetric inequality (3.16).

For the fractional kernel it also holds that

$$\frac{d}{dr} \operatorname{Per}_s(B_r) = H_{\partial B_r}^s(r, \bar{0}) \operatorname{Per}(B_r). \tag{3.18}$$

One can find a proof in [32]. Nevertheless, having in mind (3.11), a simple calculation gives

$$\frac{d}{dr} \operatorname{Per}_s(B_r) = \frac{N-s}{N\omega_N} \frac{1}{r^s} \operatorname{Per}_s(B_1) \operatorname{Per}(B_r),$$

and, since

$$H_{\partial B_r}^s(r, \bar{0}) = \frac{1}{r^s} H_{\partial B_1}^s(1, \bar{0}),$$

one gets easily that

$$\frac{d}{dr} \operatorname{Per}_s(B_r) = \frac{N-s}{N\omega_N} \frac{\operatorname{Per}_s(B_1)}{H_{\partial B_1}^s(1, \bar{0})} H_{\partial B_r}^s(r, \bar{0}) \operatorname{Per}(B_r).$$

And the following relation holds true for the fractional perimeter and fractional curvature of the unit ball:

$$\frac{N-s}{N\omega_N}\frac{\mathrm{Per}_s(B_1)}{H^s_{\partial B_1}(1,\overline{0})} = 1,$$

or

$$\frac{\mathrm{Per}_s(B_1)}{|B_1|} = \frac{N}{N-s}H^s_{\partial B_1}(1,\overline{0}).$$

The curvatures $H^s_{\partial B_1}(1,\overline{0})$ and $\mathrm{Per}_s(B_1)$ are dimensional constants depending on N and s. In [47] (see also [45]), one can find different descriptions of $\mathrm{Per}_s(B_1)$.

Integrating in (3.18),

$$\mathrm{Per}_s(B_r) = N\omega_N \int_0^r \tau^{N-1} H^s_{\partial B_\tau}(\tau,\overline{0})d\tau$$

$$= \int_{B_r} H^s_{\partial B_{\|x\|}}(x)dx.$$

Hence, the fractional isoperimetric inequality (3.10) can also be written as

$$\left(\frac{1}{|B_{\rho_E}|}\int_{B_{\rho_E}} H^s_{\partial B_{\|x\|}}(x)dx\right)|E| \le \mathrm{Per}_s(E). \qquad (3.19)$$

Therefore, all the three isoperimetric inequalities have a common formulation: (3.15), (3.16), and (3.19).

Observe that an explicit power of $|E|$ appears in (3.12), like in (3.9). Nevertheless, an explicit calculus of

$$\int_{B_{\rho_E}} H^J_{\partial B_{\|x\|}}(x)dx \qquad (3.20)$$

as a function of ρ_E is not clear even for the uniform J on a ball; observe that, for the J-curvature,

$$H^J_{\partial B_r}(r,\overline{0}) = H^{J_{\frac{1}{r}}}_{\partial B_1}(1,\overline{0}),$$

and hence (3.20) cannot be written in a so clean way.

Remark 3.12 In the local case it is well-known that in (3.8) we have equality if and only if E is a ball. The same happens for the fractional isoperimetric inequality (3.12). In the case of radially non-increasing J having compact support B_R, as a consequence of [24, Theorem 1], the equality in (3.13) holds if and only if D is a ball of radius r, when $r > \frac{R}{2}$. In Example 5.8 we show that the J-isoperimetric inequality is an equality for sets that are not balls.

Nonlocal Operators

4

4.1 A Characterization of the Nonlocal Perimeter

Following Gilboa–Osher [50] (see also [14]), we introduce the following nonlocal operators. For a function $u : \mathbb{R}^N \to \mathbb{R}$, we define its *nonlocal gradient* as the function $\nabla_J u : \mathbb{R}^N \times \mathbb{R}^N \to \mathbb{R}$ defined by:

$$(\nabla_J u)(x, y) = J(x - y)(u(y) - u(x)), \qquad x, y \in \mathbb{R}^N.$$

And for a function $\mathbf{z} : \mathbb{R}^N \times \mathbb{R}^N \to \mathbb{R}$, its *nonlocal divergence* $\operatorname{div}_J \mathbf{z} : \mathbb{R}^N \to \mathbb{R}$ is defined as:

$$(\operatorname{div}_J \mathbf{z})(x) = \frac{1}{2} \int_{\mathbb{R}^N} (\mathbf{z}(x, y) - \mathbf{z}(y, x)) J(x - y) dy.$$

Remark 4.1 (An Interpretation of $\operatorname{div}_J \mathbf{z}$) Suppose that $\mathbb{R}^N$ represents a continuous network, with a source uniformly distributed in $\Omega \subset \mathbb{R}^N$ and being $\mathbb{R}^N \setminus \Omega$ a sink. If the transportation activity is described by $\mathbf{z}$, in such a way that at each point $x \in \Omega$, $\mathbf{z}(x, y) J(x - y)$ is the incoming quantity of flow from $y \in \mathbb{R}^N$ and $\mathbf{z}(y, x) J(x - y)$ is the outcoming flow at $y \in \mathbb{R}^N$, then

$$2(\operatorname{div}_J \mathbf{z})(x) = \int_{\mathbb{R}^N} (\mathbf{z}(x, y) - \mathbf{z}(y, x)) J(x - y) dy,$$

represents the total flow at x.

© Springer Nature Switzerland AG 2019
J. M. Mazón et al., *Nonlocal Perimeter, Curvature and Minimal Surfaces for Measurable Sets*, Frontiers in Mathematics,
https://doi.org/10.1007/978-3-030-06243-9_4

For $p \geq 1$, we define the space:

$$X_J^p(\mathbb{R}^N) = \left\{ \mathbf{z} \in L^\infty(\mathbb{R}^N \times \mathbb{R}^N) \; : \; \mathrm{div}_J \mathbf{z} \in L^p(\mathbb{R}^N) \right\}.$$

Observe that $X_J^\infty(\mathbb{R}^N) = L^\infty(\mathbb{R}^N \times \mathbb{R}^N)$. For $u \in BV_J(\mathbb{R}^N) \cap L^{p'}(\mathbb{R}^N)$ and $\mathbf{z} \in X_J^p(\mathbb{R}^N)$, $1 \leq p \leq \infty$, we have the following *Green formula*:

$$\int_{\mathbb{R}^N} u(x)(\mathrm{div}_J \mathbf{z})(x)dx = -\frac{1}{2} \int_{\mathbb{R}^N \times \mathbb{R}^N} (\nabla_J u)(x, y)\mathbf{z}(x, y)dxdy. \tag{4.1}$$

In the next result, we characterize $\mathcal{TV}_J$ and the nonlocal perimeter using the nonlocal divergence operator. Let us denote by $\mathrm{sign}_0(r)$ the usual sign function and by $\mathrm{sign}(r)$ the multivalued sign function for which $\mathrm{sign}(0) = [-1, 1]$.

Proposition 4.2 *Let* $1 \leq p \leq \infty$. *For* $u \in BV_J(\mathbb{R}^N) \cap L^{p'}(\mathbb{R}^N)$, *we have*

$$\mathcal{TV}_J(u) = \sup \left\{ \int_{\mathbb{R}^N} u(x)(\mathrm{div}_J \mathbf{z})(x)dx \; : \; \mathbf{z} \in X_J^p(\mathbb{R}^N), \; \|\mathbf{z}\|_\infty \leq 1 \right\}. \tag{4.2}$$

If $u \notin BV_J(\mathbb{R}^N)$, *then*

$$\sup \left\{ \int_{\mathbb{R}^N} u(x)(\mathrm{div}_J \mathbf{z})(x)dx \; : \; \mathbf{z} \in X_J^p(\mathbb{R}^N), \; \|\mathbf{z}\|_\infty \leq 1 \right\} = +\infty.$$

In particular, for any measurable set $E \subset \mathbb{R}^N$, *we have*

$$P_J(E) = \sup \left\{ \int_E (\mathrm{div}_J \mathbf{z})(x)dx \; : \; \mathbf{z} \in X_J^1(\mathbb{R}^N), \; \|\mathbf{z}\|_\infty \leq 1 \right\}.$$

Proof Let $u \in L^1(\mathbb{R}^N) \cap L^{p'}(\mathbb{R}^N)$. Given $\mathbf{z} \in X_J^p(\mathbb{R}^N)$ with $\|\mathbf{z}\|_\infty \leq 1$, applying Green formula (4.1), we have

$$\int_{\mathbb{R}^N} u(x)(\mathrm{div}_J \mathbf{z})(x)dx = \frac{1}{2} \int_{\mathbb{R}^N \times \mathbb{R}^N} (\nabla_J u)(x, y)\mathbf{z}(x, y)dxdy$$

$$\leq \frac{1}{2} \int_{\mathbb{R}^N \times \mathbb{R}^N} |u(y) - u(x)| J(x - y)dxdy$$

$$= \mathcal{TV}_J(u).$$

Therefore,

$$\sup \left\{ \int_{\mathbb{R}^N} u(x)(\mathrm{div}_J \mathbf{z})(x)dx \; : \; \mathbf{z} \in X_J^p(\mathbb{R}^N), \; \|\mathbf{z}\|_\infty \leq 1 \right\} \leq \mathcal{TV}_J(u).$$

On the other hand, if we define

$$\mathbf{z}_n(x, y) = \text{sign}_0(u(y) - u(x))\chi_{B_n(0,0)}(x, y),$$

then $\mathbf{z}_n \in X_J^P(\mathbb{R}^N)$ with $\|\mathbf{z}_n\|_\infty \leq 1$ and

$$\mathcal{TV}_J(u) = \frac{1}{2} \int_{\mathbb{R}^N \times \mathbb{R}^N} |u(y) - u(x)| J(x - y) dx dy$$

$$= \lim_{n \to \infty} \frac{1}{2} \int_{B_n(0,0)} |u(y) - u(x)| J(x - y) dx dy$$

$$= \lim_{n \to \infty} \frac{1}{2} \int_{\mathbb{R}^N \times \mathbb{R}^N} (\nabla_J u)(x, y) \mathbf{z}_n(x, y) dx dy$$

$$= \lim_{n \to \infty} \int_{\mathbb{R}^N} u(x)(\text{div}_J \mathbf{z}_n)(x) dx$$

$$\leq \sup \left\{ \int_{\mathbb{R}^N} u(x)(\text{div}_J \mathbf{z})(x) dx \ : \ \mathbf{z} \in X_J^P(\mathbb{R}^N), \ \|\mathbf{z}\|_\infty \leq 1 \right\}.$$

$\square$

Corollary 4.3 *For $u \in L^1(\mathbb{R}^N)$,*

$$\mathcal{TV}_J(u) = \sup \left\{ \int_{\mathbb{R}^N} u(x)(\text{div}_J \mathbf{z})(x) dx \ : \ \mathbf{z} \in L^\infty(\mathbb{R}^N \times \mathbb{R}^N), \ \|\mathbf{z}\|_\infty \leq 1 \right\}.$$

In particular, for any measurable set $E \subset \mathbb{R}^N$ with finite measure,

$$P_J(E) = \sup \left\{ \int_E (\text{div}_J \mathbf{z})(x) dx \ : \ \mathbf{z} \in L^\infty(\mathbb{R}^N \times \mathbb{R}^N), \ \|\mathbf{z}\|_\infty \leq 1 \right\}. \tag{4.3}$$

4.2 Nonlocal 1-Laplacian

Nonlocal evolution equations of the form:

$$u_t(x, t) = \int_{\mathbb{R}^N} J(x - y) |u(y, t) - u(x, t)|^{P-2} (u(y, t) - u(x, t)) dy, \qquad p \geq 1$$

and variations of it, have been recently widely used to model diffusion processes (see [12] and the references therein).

For $p = 1$, using the nonlocal calculus from Sect. 4.1, we have formally

$$\operatorname{div}_J \left(\frac{\nabla_J u}{|\nabla_J u|} \right)(x) = \frac{1}{2} \int_{\mathbb{R}^N} J(x - y) \left(\frac{\nabla_J u}{|\nabla_J u|}(x, y) - \frac{\nabla_J u}{|\nabla_J u|}(y, x) \right) dy$$

$$= \frac{1}{2} \int_{\mathbb{R}^N} J(x - y) \left(\frac{(u(y) - u(x))J(x - y) - (u(x) - u(y))J(x - y)}{J(x - y)|u(y) - u(x)|} \right) dy$$

$$= \int_{\mathbb{R}^N} J(x - y) \frac{u(y) - u(x)}{|u(y) - u(x)|} dy,$$

that we have called *nonlocal 1-Laplacian operator* in [10] and [11]:

$$\Delta_1^J u(x) = \int_{\mathbb{R}^N} J(x - y) \frac{u(y) - u(x)}{|u(y) - u(x)|} dy \qquad \text{for } u \in L^1(\mathbb{R}^N), \ x \in \mathbb{R}^N.$$

Also, if $p > 1$, we have

$$\operatorname{div}_{J^{1/p}} \left(|\nabla_{J^{1/p}} u|^{p-2} \nabla_{J^{1/p}} u \right)(x) = \int_{\mathbb{R}^N} J(x - y)|u(y) - u(x)|^{p-2}(u(y) - u(x))dy,$$

that we have called *nonlocal p-Laplacian operator*. In [10] and [11], we have studied these nonlocal operators with different boundary conditions, and from these works we take the following definition.

Definition 4.4 Given $v \in L^1(\mathbb{R}^N)$, we say that $u \in L^1(\mathbb{R}^N)$ is a solution of

$$-\Delta_1^J u \ni v \quad \text{in } \mathbb{R}^N$$

if there exists $\mathbf{g} \in L^\infty(\mathbb{R}^N \times \mathbb{R}^N)$ with $\|g\|_\infty \leq 1$ verifying

$$\mathbf{g}(x, y) = -\mathbf{g}(y, x) \quad \text{for } (x, y) \text{ a.e in } \mathbb{R}^N \times \mathbb{R}^N,$$

$$J(x - y)\mathbf{g}(x, y) \in J(x - y)\operatorname{sign}(u(y) - u(x)) \quad \text{a.e } (x, y) \in \mathbb{R}^N \times \mathbb{R}^N,$$

and

$$-\int_{\mathbb{R}^N} J(x - y)\mathbf{g}(x, y)\, dy = v(x) \quad \text{a.e } x \in \mathbb{R}^N.$$

We point out that, in general, the operator Δ_1^J is multivalued (see Remark 5.11).

In order to study the Cauchy problem associated with the nonlocal 1-Laplacian, we will see that we can consider it as the gradient flow in $L^2(\mathbb{R}^N)$ of the functional $\mathcal{TV}_J$. For that, we consider now the functional:

$$\mathcal{F}^J : L^2(\mathbb{R}^N) \to]-\infty, +\infty]$$

defined by:

$$\mathcal{F}^J(u) = \begin{cases} \mathcal{TV}_J(u) & \text{if } u \in L^2(\mathbb{R}^N) \cap BV_J(\mathbb{R}^N), \\ +\infty & \text{if } u \in L^2(\mathbb{R}^N) \setminus BV_J(\mathbb{R}^N), \end{cases}$$

which is convex and lower semicontinuous. Following the method used in [9] to get the characterization of the subdifferential of the total variation, we get the following characterization of the subdifferential of the functional $\mathcal{F}^J$.

Given a functional $\Phi : L^2(\mathbb{R}^N) \to [0, \infty]$, we define $\widetilde{\Phi} : L^2(\mathbb{R}^N) \to [0, \infty]$ as:

$$\widetilde{\Phi}(v) = \sup \left\{ \frac{\displaystyle\int_{\mathbb{R}^N} v(x)w(x)dx}{\Phi(w)} : w \in L^2(\mathbb{R}^N) \right\}$$

with the convention that $\frac{0}{0} = \frac{0}{\infty} = 0$. Obviously, if $\Phi_1 \le \Phi_2$, then $\widetilde{\Phi}_2 \le \widetilde{\Phi}_1$.

Theorem 4.5 *Let $u \in L^1(\mathbb{R}^N) \cap L^2(\mathbb{R}^N)$ and $v \in L^2(\mathbb{R}^N)$. The following assertions are equivalent:*

(i) $v \in \partial \mathcal{F}^J(u)$;
(ii) There exists $\mathbf{z} \in X_J^2(\mathbb{R}^N)$, $\|\mathbf{z}\|_\infty \le 1$, such that

$$v = -\mathrm{div}_J \mathbf{z} \tag{4.4}$$

and

$$\int_{\mathbb{R}^N} u(x)v(x)dx = \mathcal{F}^J(u);$$

(iii) There exists $\mathbf{z} \in X_J^2(\mathbb{R}^N)$, $\|\mathbf{z}\|_\infty \le 1$, such that (4.4) holds and

$$\mathcal{F}^J(u) = \frac{1}{2} \int_{\mathbb{R}^N \times \mathbb{R}^N} \nabla_J u(x, y)\mathbf{z}(x, y)dxdy;$$

(iv) $-\Delta_1^J u \ni v$ *in* $\mathbb{R}^N$;
 (v) There exists $\mathbf{g} \in L^\infty(\mathbb{R}^N \times \mathbb{R}^N)$ *antisymmetric with* $\|\mathbf{g}\|_\infty \le 1$ *such that*

$$- \int_{\mathbb{R}^N} J(x-y)\mathbf{g}(x,y)\,dy = v(x) \quad a.e\ x \in \mathbb{R}^N, \tag{4.5}$$

and

$$- \int_{\mathbb{R}^N} \int_{\mathbb{R}^N} J(x-y)\mathbf{g}(x,y)dy\,u(x)dx = \mathcal{F}^J(u).$$

Proof Since $\mathcal{F}^J$ is convex, lower semicontinuous and positive homogeneous of degree 1, by Andreu et al. [9, Theorem 1.8], we have

$$\partial \mathcal{F}^J(u) = \left\{ v \in L^2(\mathbb{R}^N) \ : \ \widetilde{\mathcal{F}^J}(v) \le 1, \ \int_{\mathbb{R}^N} u(x)v(x)dx = \mathcal{F}^J(u) \right\}. \tag{4.6}$$

We define, for $v \in L^2(\mathbb{R}^N)$,

$$\Psi(v) = \inf \left\{ \|\mathbf{z}\|_\infty \ : \ \mathbf{z} \in X_J^2(\mathbb{R}^N),\ v = -\mathrm{div}_J \mathbf{z} \right\}. \tag{4.7}$$

Observe that Ψ is convex, lower semicontinuous and positive homogeneous of degree 1. Moreover, it is easy to see that, if $\Psi(v) < \infty$, the infimum in (4.7) is attained, i.e. there exists some $\mathbf{z} \in X_J^2(\mathbb{R}^N)$, $v = -\mathrm{div}_J \mathbf{z}$ and $\Psi(v) = \|\mathbf{z}\|_\infty$.

Let us see that

$$\Psi = \widetilde{\mathcal{F}^J}.$$

If $\Psi(v) = \infty$, then we have $\widetilde{\mathcal{F}^J}(v) \le \Psi(v)$. Thus, we may assume that $\Psi(v) < \infty$. Let $\mathbf{z} \in L^\infty(\mathbb{R}^N \times \mathbb{R}^N)$ such that $v = -\mathrm{div}_J \mathbf{z}$. Then, for $w \in L^2(\mathbb{R}^N)$, we have

$$\int_{\mathbb{R}^N} w(x)v(x)dx = \frac{1}{2} \int_{\mathbb{R}^N \times \mathbb{R}^N} (\nabla_J w)(x,y)\mathbf{z}(x,y)dxdy \le \|\mathbf{z}\|_\infty \mathcal{F}^J(w).$$

Taking supremum in w, we obtain that $\widetilde{\mathcal{F}^J}(v) \le \|\mathbf{z}\|_\infty$. Now, taking infimum in $\mathbf{z}$, we get $\widetilde{\mathcal{F}^J}(v) \le \Psi(v)$.

To prove the opposite inequality, let us denote

$$D = \{\mathrm{div}_J \mathbf{z} \ : \ \mathbf{z} \in X_J^2(\mathbb{R}^N)\}.$$

Then, by (4.2), we have, for $v \in L^2(\mathbb{R}^N)$,

$$\tilde{\Psi}(v) = \sup \left\{ \frac{\int_{\mathbb{R}^N} w(x)v(x)dx}{\Psi(w)} \; : \; w \in L^2(\mathbb{R}^N) \right\}$$

$$\geq \sup \left\{ \frac{\int_{\mathbb{R}^N} w(x)v(x)dx}{\Psi(w)} \; : \; w \in D \right\}$$

$$= \sup \left\{ \frac{\int_{\mathbb{R}^N} \operatorname{div}_J \mathbf{z}(x)v(x)dx}{\|\mathbf{z}\|_\infty} \; : \; \mathbf{z} \in X_J^2(\mathbb{R}^N) \right\}$$

$$= \mathcal{F}^J(v).$$

Observe that the last term is equal to $+\infty$ if $v \in L^2(\mathbb{R}^N) \setminus BV_J(\mathbb{R}^N)$. Thus, $\mathcal{F}^J \leq \tilde{\Psi}$, which implies by Andreu et al. [9, Proposition 1.6], that $\Psi = \tilde{\tilde{\Psi}} \leq \tilde{\mathcal{F}^J}$. Therefore, $\Psi = \tilde{\mathcal{F}^J}$, and consequently from (4.6), we get

$$\partial \mathcal{F}^J(u) = \left\{ v \in L^2(\mathbb{R}^N) \; : \; \Psi(v) \leq 1, \; \int_{\mathbb{R}^N} u(x)v(x)dx = \mathcal{F}_1^J(u) \right\}$$

$$= \left\{ v \in L^2(\mathbb{R}^N) \; : \; \exists \mathbf{z} \in X_J^2(\mathbb{R}^N), \; v = -\operatorname{div}_J \mathbf{z}, \; \|\mathbf{z}\|_\infty \leq 1, \right.$$

$$\left. \int_{\mathbb{R}^N} u(x)v(x)dx = \mathcal{F}^J(u) \right\},$$

from where it follows the equivalence between (i) and (ii).

To get the equivalence between (ii) and (iii), we only need to apply the Green formula (4.1).

By the antisymmetry of $\mathbf{g}$, it is easy to see that (iv) and (v) are equivalent. On the other hand, to see that (iii) implies (v), it is enough to take $\mathbf{g}(x, y) = \frac{1}{2}(\mathbf{z}(x, y) - \mathbf{z}(y, x))$. Finally, to see that (v) implies (ii), it is enough to take $\mathbf{z}(x, y) = \mathbf{g}(x, y)$ (observe that, from (4.5), $-\operatorname{div}_J(\mathbf{g}) = v$, so $\mathbf{g} \in X_J^2(\mathbb{R}^N)$). $\qquad \square$

Remark 4.6 Observe that if $v \in \partial \mathcal{F}^J(u)$, then any function $\mathbf{z}$ satisfying the characterization of Theorem 4.5 also satisfies

$$J(x - y)\mathbf{z}(x, y) \in J(x - y)\operatorname{sign}((u(y) - u(x))) \quad \text{a.e. in } \mathbb{R}^N \times \mathbb{R}^N.$$

But, moreover, we can choose one being antisymmetric.

By Theorem 4.5 and following [12, Theorem 7.5], it is easy to prove the following result.

Lemma 4.7 $\partial \mathcal{F}^J$ *is an m-completely accretive operator in* $L^2(\mathbb{R}^N)$.

As consequence of Theorem 4.5 and Lemma 4.7, we can give the following existence and uniqueness result for the Cauchy problem:

$$
\begin{cases}
u_t - \Delta_1^J u \ni 0 & \text{in } (0, T) \times \mathbb{R}^N \\
u(0, x) = u_0(x) \ x \in \mathbb{R}^N.
\end{cases}
\tag{4.8}
$$

Theorem 4.8 *For every* $u_0 \in L^2(\mathbb{R}^N)$, *there exists a unique solution of the Cauchy problem* (4.8) *in* $(0, T)$ *for any* $T > 0$, *in the following sense: the solution* $u \in W^{1,1}(0, T; L^2(\mathbb{R}^N))$, $u(0, \cdot) = u_0$, *and for almost all* $t \in (0, T)$

$$
u_t(\cdot, t) - \Delta_1^J u(t) \ni 0.
$$

Moreover, we have the following contraction principle in any $L^q(\mathbb{R}^N)$-*space with* $1 \le q \le \infty$:

$$
\|u(t) - v(t)\|_q \le \|u_0 - v_0\|_q \quad \forall 0 < t < T,
$$

for any pair of solutions, u, v, *of problem* (4.8) *with initial data* u_0, v_0, *respectively.*

Proof By the theory of maximal monotone operators (see [19]), and having in mind the characterization of the subdifferential of $\mathcal{F}^J$, for every $u_0 \in L^2(\Omega)$, there exists a unique strong solution of the abstract Cauchy problem:

$$
\begin{cases}
u'(t) + \partial \mathcal{F}^J(u(t)) \ni 0, & t \in (0, T), \\
u(0) = u_0,
\end{cases}
$$

that is exactly the concept of solution given. The contraction principle is consequence of being the operator completely accretive (see [16]). $\qquad \square$

Nonlocal Cheeger and Calibrable Sets

5.1   Nonlocal Cheeger Sets

Given a non-null, measurable and bounded set $\Omega \subset \mathbb{R}^N$, we define its *J-Cheeger constant* by:

$$h_1^J(\Omega) = \inf \left\{ \frac{P_J(E)}{|E|} : E \subset \Omega, \ E \text{ measurable with } |E| > 0 \right\}. \tag{5.1}$$

As a consequence of (1.14), we have that $\frac{P_J(E)}{|E|} \leq 1$. Hence,

$$h_1^J(\Omega) \leq 1.$$

A measurable set $E_\Omega \subset \Omega$ achieving the infimum in (5.1) is said to be a *J-Cheeger set* of Ω.

To get a lower bound for $h_1^J(\Omega)$, we recall the following Poincaré-type inequality given in [12, Proposition 6.25].

Proposition 5.1 *Suppose J is continuous. Given Ω a bounded domain in $\mathbb{R}^N$, $p \geq 1$ and $\psi \in L^p(\Omega_J \setminus \overline{\Omega})$, there exists $\lambda(J, \Omega, p) > 0$ such that*

$$\lambda(J, \Omega, p) \int_\Omega |u(x)|^p \, dx \leq \int_\Omega \int_{\Omega_J} J(x-y)|u_\psi(y) - u(x)|^p \, dy \, dx + \int_{\Omega_J \setminus \overline{\Omega}} |\psi(y)|^p \, dy$$

for all $u \in L^p(\Omega)$.

© Springer Nature Switzerland AG 2019
J. M. Mazón et al., *Nonlocal Perimeter, Curvature and Minimal Surfaces for Measurable Sets*, Frontiers in Mathematics,
https://doi.org/10.1007/978-3-030-06243-9_5

Proof First, let us assume that there exist $r, \alpha > 0$ such that $J(x) \geq \alpha$ in B_r. Let

$$\mathcal{O}_0 = \{x \in \Omega_J \setminus \overline{\Omega} \ : \ d(x, \Omega) \leq r/2\},$$

$$\mathcal{O}_1 = \{x \in \Omega \ : \ d(x, \mathcal{O}_0) \leq r/2\},$$

$$\mathcal{O}_j = \left\{ x \in \Omega \setminus \cup_{k=1}^{j-1} \mathcal{O}_k \ : \ d(x, \mathcal{O}_{j-1}) \leq r/2 \right\}, \quad j = 2, 3, \ldots$$

Observe that we can cover Ω by a finite number of non-null sets $\{\mathcal{O}_j\}_{j=1}^{l_r}$. Now,

$$\int_\Omega \int_{\Omega_J} J(x-y)|u_\psi(y) - u(x)|^p \, dy \, dx \geq \int_{\mathcal{O}_j} \int_{\mathcal{O}_{j-1}} J(x-y)|u_\psi(y) - u(x)|^p \, dy \, dx,$$

$j = 1, \ldots, l_r$, and

$$\int_{\mathcal{O}_j} \int_{\mathcal{O}_{j-1}} J(x-y)|u_\psi(y) - u(x)|^p \, dy \, dx$$

$$\geq \frac{1}{2^p} \int_{\mathcal{O}_j} \int_{\mathcal{O}_{j-1}} J(x-y)|u(x)|^p \, dy \, dx$$

$$- \int_{\mathcal{O}_j} \int_{\mathcal{O}_{j-1}} J(x-y)|u_\psi(y)|^p \, dy \, dx$$

$$= \frac{1}{2^p} \int_{\mathcal{O}_j} \left(\int_{\mathcal{O}_{j-1}} J(x-y) \, dy \right) |u(x)|^p \, dx$$

$$- \int_{\mathcal{O}_{j-1}} \left(\int_{\mathcal{O}_j} J(x-y) \, dx \right) |u_\psi(y)|^p \, dy$$

$$\geq \frac{1}{2^p} \min_{x \in \overline{\mathcal{O}_j}} \int_{\mathcal{O}_{j-1}} J(x-y) dy \int_{\mathcal{O}_j} |u(x)|^p \, dx$$

$$- \int_{\mathcal{O}_{j-1}} |u_\psi(y)|^p \, dy,$$

since $\int_{\mathbb{R}^N} J(x)dx = 1$. Hence,

$$\int_\Omega \int_{\Omega_J} J(x-y)|u_\psi(y) - u(x)|^p \, dy \, dx \geq \alpha_j \int_{\mathcal{O}_j} |u(x)|^p \, dx - \int_{\mathcal{O}_{j-1}} |u_\psi(y)|^p \, dy,$$

where

$$\alpha_j = \frac{1}{2^p} \min_{x \in \overline{\mathcal{O}}_j} \int_{\mathcal{O}_{j-1}} J(x-y) dy > 0.$$

Therefore, since $u_\psi(y) = \psi(y)$ if $y \in \mathcal{O}_0$, $u_\psi(y) = u(y)$ if $y \in \mathcal{O}_j$, $j = 1, \ldots, l_r$, $\mathcal{O}_j \cap \mathcal{O}_i = \emptyset$, for all $i \neq j$ and $|\Omega \setminus \cup_{j=1}^{j_r} \mathcal{O}_j| = 0$, it is easy to see that there exists $\hat{\lambda} = \hat{\lambda}(J, \Omega, p) > 0$ such that

$$\int_\Omega |u|^p \leq \hat{\lambda} \int_\Omega \int_{\Omega_J} J(x-y)|u_\psi(y) - u(x)|^p \, dy \, dx + \hat{\lambda} \int_{\mathcal{O}_0} |\psi|^p.$$

The proof is finished by taking $\lambda(J, \Omega, p) = \hat{\lambda}^{-1}$.

In the general case, we have that there exist a ≥ 0 and $r, \alpha > 0$ such that

$$J(x) \geq \alpha \text{ in the annulus } A(0, \text{a}, r). \tag{5.2}$$

In this case, we proceed as before with the same choice of the sets $\mathcal{O}_j$ for $j \geq 0$ and

$$\mathcal{O}_{-j} = \left\{ x \in \Omega_J \setminus (\Omega \cup \cup_{k=0}^{j-1} \mathcal{O}_{-k}) : d(x, \mathcal{O}_{-j+1}) \leq r/2 \right\}, \quad j = 1, 2, 3, \ldots$$

Observe that for each $\mathcal{O}_j$, $j \geq 1$, there exists $\mathcal{O}_{j^e}$ with $j^e < j$ and such that

$$|(x + A(0, \text{a}, r)) \cap \mathcal{O}_{j^e}| > 0 \quad \forall x \in \overline{\mathcal{O}}_j. \tag{5.3}$$

With this choice of $\mathcal{O}_j$ and taking into account (5.2) and (5.3), as before, we obtain

$$\int_\Omega \int_{\Omega_J} J(x-y)|u_\psi(y) - u(x)|^p \, dy \, dx \geq \int_{\mathcal{O}_j} \int_{\mathcal{O}_{j^e}} J(x-y)|u_\psi(y) - u(x)|^p \, dy \, dx$$

$$\geq \alpha_j \int_{\mathcal{O}_j} |u(x)|^p \, dx - \int_{\mathcal{O}_{j^e}} |u_\psi(y)|^p \, dy,$$

$j = 1, \ldots, l_r$, where

$$\alpha_j = \frac{1}{2^p} \min_{x \in \overline{\mathcal{O}}_j} \int_{\mathcal{O}_{j^e}} J(x-y) dy > 0.$$

And, we conclude as before.   □

Proposition 5.2 *Suppose J is continuous. Then,*

$$h_1^J(\Omega) \geq \frac{\lambda(J, \Omega, p)}{2},$$

where $\lambda(J, \Omega, p)$ is given in Proposition 5.1.

Proof From Proposition 5.1, for $p = 1$ and $\psi = 0$, we get that

$$\lambda(J, \Omega, 1)|E| \leq 2P_J(E) \quad \text{for all measurable } E \subset \Omega,$$

and consequently,

$$h_1^J(\Omega) \geq \frac{\lambda(J, \Omega, 1)}{2}.$$

$\square$

It is well known (see [49]) that the classical Cheeger constant:

$$h_1(\Omega) = \inf\left\{ \frac{\text{Per}(E)}{|E|} \; : \; E \subset \Omega, \; |E| > 0 \right\},$$

for Ω a bounded smooth domain, is an optimal Poincaré constant, namely, it coincides with the first eigenvalue of the 1-Laplacian:

$$h_1(\Omega) = \inf\left\{ \frac{\displaystyle\int_\Omega |Du| + \int_{\partial\Omega} |u| d\mathcal{H}^{N-1}}{\|u\|_{L^1(\Omega)}} \; : \; u \in BV(\Omega), \; u \neq 0 \right\}.$$

It is also characterized (see [52]) as:

$$h_1(\Omega) = \sup\left\{ h \in \mathbb{R} \; : \; \exists V \in L^\infty(\Omega, \mathbb{R}^N), \; \|V\|_\infty \leq 1, \; \text{div} V \geq h \right\},$$

which is usually referred to as a continuous version of the Min Cut Max Flow Theorem.

In the next result, we obtain nonlocal versions of these two characterizations of the J-Cheeger constant given in (5.1).

Theorem 5.3 *Let Ω be a non-null, measurable and bounded set of $\mathbb{R}^N$, then*

$$h_1^J(\Omega) = \inf\left\{\mathcal{TV}_J(u) \ : \ u \in BV_J(\mathbb{R}^N), \ u = 0 \text{ in } \mathbb{R}^N \setminus \Omega, \ \|u\|_1 = 1\right\}$$

$$= \inf\left\{\frac{\mathcal{TV}_J(u)}{\|u\|_1} \ : \ u \in BV_J(\mathbb{R}^N), \ u = 0 \text{ in } \mathbb{R}^N \setminus \Omega, \ u \neq 0\right\}, \tag{5.4}$$

and

$$h_1^J(\Omega) = \sup\left\{h \in \mathbb{R}^+ \ : \ \exists \mathbf{z} \in X_J^\infty(\mathbb{R}^N), \ \|\mathbf{z}\|_\infty \leq 1, \ \operatorname{div}_J \mathbf{z} \geq h \text{ in } \Omega\right\}$$

$$= \sup\left\{\frac{1}{\|\mathbf{z}\|_\infty} \ : \ \operatorname{div}_J \mathbf{z} = \chi_\Omega\right\} \tag{5.5}$$

$$= \sup\left\{\frac{1}{\|\mathbf{z}\|_\infty} \ : \ \operatorname{div}_J \mathbf{z} = 1 \text{ in } \Omega\right\}.$$

Proof Given a measurable subset $E \subset \Omega$ with $|E| > 0$, we have

$$\frac{\mathcal{TV}_J(\chi_E)}{\|\chi_E\|_1} = \frac{P_J(E)}{|E|}.$$

Therefore,

$$\inf\left\{\frac{\mathcal{TV}_J(u)}{\|u\|_1} \ : \ u \in BV_J(\mathbb{R}^N), \ u = 0 \text{ in } \mathbb{R}^N \setminus \Omega, \ u \neq 0\right\} \leq h_1^J(\Omega).$$

On the other hand, by the coarea formula (2.19) and Cavalieri's formula, given $u \in BV_J(\mathbb{R}^N)$, with $u = 0$ in $\mathbb{R}^N \setminus \Omega$ and $u \neq 0$, we have

$$\mathcal{TV}_J(u) = \int_{-\infty}^{+\infty} P_J(E_t(u))\, dt \geq h_1^J(\Omega) \int_{-\infty}^{+\infty} |E_t(u)|\, dt = h_1^J(\Omega)\|u\|_1$$

and taking infimum we get

$$\inf\left\{\frac{\mathcal{TV}_J(u)}{\|u\|_1} \ : \ u \in BV_J(\mathbb{R}^N), \ u = 0 \text{ in } \mathbb{R}^N \setminus \Omega, \ u \neq 0\right\} \geq h_1^J(\Omega).$$

Therefore, (5.4) holds true.

Let

$$A = \left\{h \in \mathbb{R}^+ \ : \ \exists \mathbf{z} \in X_J^\infty(\mathbb{R}^N), \ \|\mathbf{z}\|_\infty \leq 1, \ \operatorname{div}_J \mathbf{z} \geq h \text{ in } \Omega\right\}$$

and

$$\alpha = \sup A$$

(observe that $0 \le \alpha \le 1$). Given $h \in A$ and $E \subset \Omega$ with $|E| > 0$, applying (4.3), we get

$$h|E| = \int_E h \, dx \le \int_E \operatorname{div}_J \mathbf{z}(x) dx \le P_J(E).$$

Hence,

$$h \le \frac{P_J(E)}{|E|},$$

and, taking supremum in h and infimum in E, we obtain that $\alpha \le h_1^J(\Omega)$.

By (5.4), we have

$$\frac{1}{h_1^J(\Omega)} = \sup \left\{ \frac{\|u\|_1}{\mathcal{TV}_J(u)} \; : \; u \in BV_J(\mathbb{R}^N), \; u = 0 \text{ in } \mathbb{R}^N \setminus \Omega, \; u \ne 0 \right\}$$

$$= \sup \left\{ \frac{\|u\|_1}{\mathcal{TV}_J(u)} \; : \; u \in BV_J(\mathbb{R}^N), \; u \ge 0, \; u = 0 \text{ in } \mathbb{R}^N \setminus \Omega, \; u \ne 0 \right\}$$

$$= \sup \left\{ \int_\Omega u(x) dx \; : \; \mathcal{TV}_J(u) \le 1, \; u \in BV_J(\mathbb{R}^N), \; u = 0 \text{ in } \mathbb{R}^N \setminus \Omega \right\}$$

$$= \sup \left\{ \langle u, \chi_\Omega \rangle - \Xi(L(u)) \; : \; u \in L^1(\mathbb{R}^N) \right\},$$

being $L : L^1(\mathbb{R}^N) \to L^1(\mathbb{R}^N \times \mathbb{R}^N)$ the linear map:

$$L(u)(x, y) = \frac{1}{2}(u(x) - u(y)) J(x - y),$$

and $\Xi : L^1(\mathbb{R}^N \times \mathbb{R}^N) \to [0, +\infty]$ the convex function:

$$\Xi(w) = \begin{cases} 0 & \text{if } \|w\|_{L^1(\mathbb{R}^N \times \mathbb{R}^N)} \le 1, \\ +\infty & \text{otherwise.} \end{cases}$$

By the Frenchel–Rockafeller duality Theorem [21, Th. 1.12] and having in mind [40, Prop. 5], we have

$$\sup \left\{ \langle u, \chi_\Omega \rangle - \Xi(L(u)) \; : \; u \in L^1(\mathbb{R}^N) \right\} = \inf \left\{ \Xi^*(\mathbf{z}) \; : \; L^*(\mathbf{z}) = \chi_\Omega \right\}.$$

Now,

$$\Xi^*(\mathbf{z}) = \sup\left\{\int_{\mathbb{R}^N \times \mathbb{R}^N} \mathbf{z}(x,y)w(x,y)dxdy - \Xi(w) \ : \ w \in L^1(\mathbb{R}^N \times \mathbb{R}^N)\right\} = \|\mathbf{z}\|_\infty.$$

On the other hand,

$$\langle L^*(\mathbf{z}), u \rangle = \langle \mathbf{z}, L(u) \rangle = \int_{\mathbb{R}^N \times \mathbb{R}^N} \mathbf{z}(x,y) \frac{1}{2}(u(x) - u(y))J(x - y)dxdy$$

$$= \frac{1}{2}\int_{\mathbb{R}^N \times \mathbb{R}^N} (\mathbf{z}(x,y) - \mathbf{z}(y,x))J(x - y)u(x)dxdy = \langle \mathrm{div}_J \mathbf{z}, u \rangle,$$

that is, $L^*(\mathbf{z}) = \mathrm{div}_J \mathbf{z}$. Consequently,

$$\frac{1}{h_1^J(\Omega)} = \inf\{\|\mathbf{z}\|_\infty \ : \ \mathrm{div}_J \mathbf{z} = \chi_\Omega\},$$

from where it follows that

$$h_1^J(\Omega) = \sup\left\{\frac{1}{\|\mathbf{z}\|_\infty} \ : \ \mathrm{div}_J \mathbf{z} = \chi_\Omega\right\} \le \sup\left\{\frac{1}{\|\mathbf{z}\|_\infty} \ : \ \mathrm{div}_J \mathbf{z} = 1 \text{ in } \Omega\right\} \le \alpha,$$

and we finish the proof of (5.5).   □

Remark 5.4 It is well known that every bounded domain $\Omega \subset \mathbb{R}^N$ with Lipschitz boundary contains a classical Cheeger set E, that is, a set $E \subset \Omega$ such that

$$h_1(\Omega) = \frac{\mathrm{Per}(E)}{|E|}.$$

Furthermore, in [4] it is proved that there is a unique Cheeger set inside any nontrivial convex body in $\mathbb{R}^N$, being this Cheeger set convex.

On the other hand, in [18] it is proved that for any $s \in (0,1)$, every open and bounded set $\Omega \subset \mathbb{R}^N$ admits and s-Cheeger set, that is, a set $E \subset \Omega$ such that

$$\frac{\mathrm{Per}_s(E)}{|E|} = \inf\left\{\frac{\mathrm{Per}_s(F)}{|F|} \ : \ F \subset \Omega, \ |F| > 0\right\}.$$

We will show in Example 5.18 that there are convex sets without a J-Cheeger set. This is due to the lack of compactness when one considers nonsingular kernels.

5.2   Nonlocal Calibrable Sets

We say that Ω is *J-calibrable* if it is a J-Cheeger set of itself, that is, if Ω is a non-null measurable bounded set and

$$h_1^J(\Omega) = \frac{P_J(\Omega)}{|\Omega|}.$$

As a consequence of the Isoperimetric Inequality given in Theorem 2.1, we show that any ball is J-calibrable when J is radially nonincreasing.

Proposition 5.5 *Assume J is radially nonincreasing. The following properties hold true:*

1. For $0 < r < R$,

$$\frac{P_J(B_r(x_0))}{|B_r(x_0)|} > \frac{P_J(B_R(x_0))}{|B_R(x_0)|}. \tag{5.6}$$

2. Any ball $B_R(x_0) \subset \mathbb{R}^N$ is J-calibrable.

Proof We can take $x_0 = 0$ as the centre of the ball. First, we prove that a ball B_R is J-calibrable if and only if the function

$$\Theta(r) = \frac{P_J(B_r)}{|B_r|}$$

verifies that

$$\Theta(r) \geq \Theta(R), \qquad \forall r \in (0, R). \tag{5.7}$$

Obviously, the condition is necessary. On the other hand, given $E \subset B_R$, by the isoperimetric inequality we have that

$$\frac{P_J(E)}{|E|} \geq \frac{P_J(B_r)}{|B_r|} = \Theta(r)$$

where B_r is a ball such that $|B_r| = |E|$. Hence, since we are assuming that the function $\Theta(r)$ verifies (5.7), we have

$$\frac{P_J(E)}{|E|} \geq \Theta(r) \geq \Theta(R) = \frac{P_J(B_R)}{|B_R|},$$

and consequently B_R is J-calibrable.

By the above characterization, we need to show that (5.7) holds. Let us prove that in fact the inequality is strict, that is, that (5.6) holds true. From (1.11),

$$P_J(B_r) = |B_r| - \int_{B_r} \int_{B_r} J(x - y) dy dx.$$

Hence, (5.6) is true if and only if

$$F(r) < F(R) \quad \text{for every } 0 < r < R,$$

where

$$F(r) = \frac{1}{|B_r|} \int_{B_r} \left(\int_{B_r} J(x - y) dy \right) dx.$$

Take $0 < r < R$. Then, changing variables $z = \frac{R}{r} x$, we have

$$F(r) = \frac{1}{|B_r|} \int_{B_r} \left(\int_{B_r} J(x - y) dy \right) dx$$

$$= \frac{1}{|B_R|} \int_{B_R} \left(\int_{B_r} J\left(\frac{r}{R} z - y \right) dy \right) dz$$

$$< \frac{1}{|B_R|} \int_{B_R} \left(\int_{B_R} J(z - y) dy \right) dz$$

$$= F(R),$$

since, for any $z \in B_R$, $B_r(\frac{r}{R}z) \subset\subset B_R(z)$ (note that $0 \in \partial B_r(\frac{r}{R}z)$), which implies

$$\int_{B_r} J\left(\frac{r}{R} z - y \right) dy = \int_{B_r(\frac{r}{R}z)} J(y) dy$$

$$< \int_{B_R(z)} J(y) dy$$

$$= \int_{B_R} J(z - y) dy.$$

$\square$

Remark 5.6 We have that Θ is indeed continuously differentiable in $]0, +\infty[$. By Lemma 3.10,

$$\Theta'(r) = \frac{H^J_{\partial B_r}(y) \text{Per}(B_r) |B_r| - \text{Per}(B_r) P_J(B_r)}{|B_r|^2},$$

with $y \in \partial B_r$ arbitrary.

Therefore, by the characterization given in the proof of Proposition 5.5 and having in mind that clearly ess $\sup\limits_{x \in B_r} H_{\partial B_r}^J(x)$ is attained on any point of the boundary of B_r, we get

$$B_r \text{ is } J\text{-calibrable} \iff \operatorname*{ess\,sup}_{x \in B_r} H_{\partial B_r}^J(x) \leq \frac{P_J(B_r)}{|B_r|}.$$

The next result shows that any measurable and non-null set inside a ball of radius $\frac{1}{2}$ is J-calibrable when $J = \frac{1}{|B_1|} \chi_{B_1}$.

Proposition 5.7 *Let* $J = \frac{1}{|B_1|} \chi_{B_1}$. *If* $\Omega \subset B_{\frac{1}{2}}$ *with* $|\Omega| > 0$, *then* Ω *is* J-*calibrable.*

Proof Let $E \subset \Omega$ non-null. For $x \in E$,

$$E \subset B_1(x).$$

Then,

$$P_J(E) = \int_E \int_{\mathbb{R} \setminus E} J(x - y)\, dy\, dx = \frac{1}{|B_1|} \int_E \int_{\mathbb{R}^N \setminus E} \chi_{B_1}(x - y) dy dx$$

$$= \frac{1}{|B_1|} \int_E \int_{\mathbb{R}^N \setminus E} \chi_{B_1(x)}(y) dy dx$$

$$= \frac{1}{|B_1|} \int_E |B_1(x) \setminus E| dx$$

$$= \frac{1}{|B_1|} \int_E (|B_1(x)| - |E \cap B_1(x)|) dx$$

$$= |E| \left(1 - \frac{|E|}{|B_1|} \right),$$

and

$$\frac{P_J(E)}{|E|} = \left(1 - \frac{|E|}{|B_1|} \right),$$

that is decreasing with $|E|$. Hence, the Cheeger constant of Ω is given by:

$$h_1^J(\Omega) = \left(1 - \frac{|\Omega|}{|B_1|} \right) = \frac{P_J(\Omega)}{|\Omega|},$$

as we wanted to show. $\square$

Example 5.8 If $J = \frac{1}{|B_1|} \chi_{B_1}$, by Example 3.6, we have

$$\psi_{J,N}(s) = N \int_0^{(s/\omega_N)^{1/N}} r^{N-1}\left(\omega_N - 2\left|B_1(r,\bar{0}) \cap B_r\right|\right)dr,$$

where $\psi_{J,N}$ is defined in (3.14). In the case $s < \omega_N/2^N$, we have $B_1(r,\bar{0}) \cap B_r = B_r$, and therefore

$$\psi_{J,N}(s) = N\omega_N \int_0^{(s/\omega_N)^{1/N}} r^{N-1}\left(1 - 2r^N\right)dr = s\left(1 - \frac{1}{\omega_N}s\right).$$

Consequently, if $|E| \le |B_{\frac{1}{2}}|$, by (3.13), the J-isoperimetric inequality stays as

$$|E|\left(1 - \frac{1}{\omega_N}|E|\right) \le P_J(E).$$

Now by Proposition 5.7, if $E \subset B_{\frac{1}{2}}$, then E is J-calibrable and

$$P_J(E) = |E|\left(1 - \frac{1}{\omega_N}|E|\right).$$

Therefore, the J-isoperimetric inequality is an equality for sets that are not balls.

In the local case, a set $\Omega \subset \mathbb{R}^N$ is called *calibrable* if

$$\frac{\text{Per}(\Omega)}{|\Omega|} = \inf\left\{\frac{\text{Per}(E)}{|E|} : E \subset \Omega,\ E \text{ with finite perimeter},\ |E| > 0\right\}.$$

Remark 5.9 For the local usual perimeter, when Ω is the union of two disjoint intervals in $\mathbb{R}$, $\Omega = (a,b) \cup (c,d)$, then the set is calibrable if and only if the two intervals have the same length (otherwise the Cheeger set inside Ω is the bigger interval). For the nonlocal perimeter with $J = \frac{1}{2}\chi_{[-1,1]}$, we have the following facts. Assume $c - b \ge 1$.

We want to compare $\frac{P_J(\Omega)}{|\Omega|}$ with the quotient of $\frac{P_J(E)}{|E|}$ for $E \subset (a,b) \cup (c,d)$. We decompose E as $E = E_1 \cup E_2$ with $E_1 = E \cap (a,b)$ and $E_2 = E \cap (c,d)$. For the case in which the two intervals that compose Ω have the same length, we have

$$\frac{P_J(\Omega)}{|\Omega|} \le \frac{P_J(E)}{|E|}$$

iff

$$\frac{P_J((a,b))}{b-a} \le \frac{P_J(E_1) + P_J(E_2)}{|E_1| + |E_2|},$$

which is true since, if $|E_i| \neq 0$ then, by the Isoperimetric Inequality,

$$\frac{P_J((a,b))}{b-a} \leq \frac{P_J(E_i)}{|E_i|}.$$

On the other hand, if $b - a > d - c$, then

$$\frac{P_J((a,b))}{|b-a|} < \frac{P_J(\Omega)}{|\Omega|}$$

and therefore we have that Ω is not J-calibrable.

In [5], it is proved the following characterization of convex calibrable set.

Theorem 5.10 ([5]) *Given a bounded convex set $\Omega \subset \mathbb{R}^N$ of class $C^{1,1}$, the following facts are equivalent:*

(a) Ω is calibrable.
(b) χ_Ω satisfies $-\Delta_1 \chi_\Omega = \frac{Per(\Omega)}{|\Omega|} \chi_\Omega$, being $\Delta_1 u = \mathrm{div}\left(\frac{Du}{|Du|}\right)$.

When Ω is convex, these statements are also equivalent to:

(c) $(N-1)\mathrm{ess}\sup\limits_{x\in\partial\Omega} H_{\partial\Omega}(x) \leq \dfrac{Per(\Omega)}{|\Omega|}.$

We are going to study the validity of a similar result to the above theorem for the nonlocal case. In the following remark, we will introduce the main idea that is behind the proof for the nonlocal case.

Remark 5.11 Let $\Omega \subset \mathbb{R}^N$ be a Borel set and assume there exist a constant $\lambda > 0$ and a function τ with $\tau(x) = 1$ in Ω such that

$$-\lambda\tau \in \Delta_1^J \chi_\Omega \quad \text{in } \mathbb{R}^N.$$

Then, there exists $\mathbf{g} \in L^\infty(\mathbb{R}^N \times \mathbb{R}^N)$, $\mathbf{g}(x,y) = -\mathbf{g}(y,x)$ for almost all $(x,y) \in \mathbb{R}^N \times \mathbb{R}^N$, $\|\mathbf{g}\|_\infty \leq 1$, satisfying

$$\int_{\mathbb{R}^N} J(x-y)\mathbf{g}(x,y)\,dy = -\lambda\tau(x) \quad \text{a.e } x \in \mathbb{R}^N.$$

with

$$J(x-y)\mathbf{g}(x,y) \in J(x-y)\mathrm{sign}(\chi_\Omega(y) - \chi_\Omega(x)) \quad \text{a.e. } (x,y) \in \mathbb{R}^N \times \mathbb{R}^N.$$

Then,

$$\lambda|\Omega| = \int_{\mathbb{R}^N} \lambda\tau(x)\chi_\Omega(x)dx$$

$$= -\int_{\mathbb{R}^N} \left(\int_{\mathbb{R}^N} J(x-y)\mathbf{g}(x,y)\,dy\right)\chi_\Omega(x)dx$$

$$= \mathcal{TV}_J(\chi_\Omega)$$

$$= P_J(\Omega),$$

and consequently,

$$\lambda = \frac{P_J(\Omega)}{|\Omega|}.$$

On the other hand, we observe again that the operator Δ_1^J is multivalued. Let us take, for example, $J = \frac{1}{2}\chi_{[-1,1]}$. We have that

$$-f \in \Delta_1^J \chi_{]-1,1[} \iff \exists \mathbf{g} \text{ antisymmetric, } \|\mathbf{g}\|_\infty \le 1$$

satisfying

$$\int_{\mathbb{R}} J(x-y)\mathbf{g}(x,y)\,dy = -f(x) \quad \text{a.e } x \in \mathbb{R}.$$

and

$$J(x-y)\mathbf{g}(x,y) \in J(x-y)\mathrm{sign}(\chi_{]-1,1[}(y) - \chi_{]-1,1[}(x)) \quad \text{a.e. } (x,y) \in \mathbb{R} \times \mathbb{R}.$$

Then, by taking for instance

$$\mathbf{g}(x,y) = \mathrm{sign}_0(\chi_{]-1,1[}(y) - \chi_{]-1,1[}(x)),$$

we have, if $x \in [-1,1]$,

$$f(x) = \frac{1}{2}|x|,$$

and, if $x \notin [-1, 1]$,

$$
f(x) = \begin{cases}
-\dfrac{1}{2}(x+2) \text{ if } -2 \le x \le -1, \\[2mm]
-\dfrac{1}{2}(2-x) \text{ if } 1 \le x \le 2, \\[2mm]
0 \qquad\qquad \text{ if } x \le -2 \text{ or } x \ge 2.
\end{cases}
$$

But, by taking

$$
\mathbf{g}(x, y) = \begin{cases}
1 \text{ if } y \in [-1, 1], x \notin [-1, 1], \\[2mm]
-1 \text{ if } x \in [-1, 1], y \notin [-1, 1], \\[2mm]
\dfrac{1}{2} \text{ in } \{0 < y < x < 1\} \cup \{-1 < x < y < 0\}, \\[2mm]
-\dfrac{1}{2} \text{ in } \{0 < x < y < 1\} \cup \{-1 < y < x < 0\}, \\[2mm]
\dfrac{1}{2} \text{ in } \{1 < y < x\} \cup \{y < x < -1\}, \\[2mm]
-\dfrac{1}{2} \text{ in } \cup \{1 < x < y\} \cup \{x < y < -1\}, \\[2mm]
0 \text{ otherwise,}
\end{cases}
$$

we get a different but interesting representation for $\Delta_1^J \chi_{]-1,1[}$. We get that

$$
-\frac{P_J(]-1, 1[)}{|]-1, 1[|}\,\tau = \frac{1}{4}\tau \in \Delta_1^J \chi_{]-1,1[} \quad \text{in } \mathbb{R}^N
$$

with

$$
\tau(x) = \begin{cases}
1 \qquad\qquad \text{if } x \in [-1, 1], \\[2mm]
-(|x| - 2)^-, \text{ otherwise,}
\end{cases}
$$

Note that this function τ verifies that

$$
\tau = 1 \text{ in }]-1, 1[,
$$

and this gives, as we will see in the next theorem, that $]-1, 1[$ is J-calibrable. Of course, in this simple case, this was obtained previously by more elementary methods.

The next result is the nonlocal version of the fact that (a) is equivalent to (b) in Theorem 5.10.

Theorem 5.12 *Let $\Omega \subset \mathbb{R}^N$ be a non-null measurable bounded set.*

(i) Assume that $\int_\Omega J(x - y)dy \geq \alpha > 0$ for all $x \in \Omega$. If Ω is J-calibrable, then there exists a function τ equal to 1 in Ω such that

$$-\frac{P_J(\Omega)}{|\Omega|}\tau \in \Delta_1^J \chi_\Omega \quad in \ \mathbb{R}^N. \tag{5.8}$$

(ii) If there exists a function τ equal to 1 in Ω and satisfying (5.8), then Ω is J-calibrable.

Proof We first prove (ii): By hypothesis, there exists $\mathbf{g}(x, y) = -\mathbf{g}(y, x)$ for almost all $(x, y) \in \mathbb{R}^N \times \mathbb{R}^N$, $\|\mathbf{g}\|_\infty \leq 1$, satisfying

$$\int_{\mathbb{R}^N} J(x - y)\mathbf{g}(x, y)\, dy = -\frac{P_J(\Omega)}{|\Omega|}\tau(x) \quad \text{a.e } x \in \mathbb{R}^N.$$

with

$$J(x - y)\mathbf{g}(x, y) \in J(x - y)\text{sign}(\chi_\Omega(y) - \chi_\Omega(x)) \quad \text{a.e. } (x, y) \in \mathbb{R}^N \times \mathbb{R}^N,$$

and τ is a function such that

$$\tau = 1 \quad \text{in } \Omega.$$

Then, if F is a bounded measurable set, $F \subset \Omega$, we have

$$\frac{P_J(\Omega)}{|\Omega|}|F| = \frac{P_J(\Omega)}{|\Omega|}\int_{\mathbb{R}^N} \tau(x)\chi_F(x)dx$$

$$= -\int_{\mathbb{R}^N}\int_{\mathbb{R}^N} J(x - y)\mathbf{g}(x, y)\chi_F(x)\, dydx$$

$$= \frac{1}{2}\int_{\mathbb{R}^N}\int_{\mathbb{R}^N} J(x - y)\mathbf{g}(x, y)(\chi_F(y) - \chi_F(x))\, dydx$$

$$\leq P_J(F),$$

Therefore,

$$h_1^J(\Omega) = \frac{P_J(\Omega)}{|\Omega|},$$

and consequently Ω is J-calibrable.

Let us now prove (i): Let $\tilde{\mathbf{g}} \in L^{\infty}(\mathbb{R}^N \times \mathbb{R}^N)$ be defined as:

$$\tilde{\mathbf{g}}(x, y) = \begin{cases} 0 & \text{if } x \in \mathbb{R}^N \setminus \Omega, \ y \in \mathbb{R}^N \setminus \Omega, \\ -1 & \text{if } x \in \Omega, \ y \in \mathbb{R}^N \setminus \Omega, \\ 1 & \text{if } x \in \mathbb{R}^N \setminus \Omega, \ y \in \Omega, \\ \hat{g}(x) & \text{if } x, y \in \Omega, \end{cases}$$

being $\hat{g}$ a function to be determined. We define

$$\tau(x) = -\frac{1}{\frac{P_J(\Omega)}{|\Omega|}} \int_{\mathbb{R}^N} J(x - y)\tilde{\mathbf{g}}(x, y)dy, \quad x \in \mathbb{R}^N.$$

For $x \in \Omega$, we have

$$\tau(x) = \frac{|\Omega|}{P_J(\Omega)} \int_{\mathbb{R}^N \setminus \Omega} J(x - y)dy - \hat{g}(x)\frac{|\Omega|}{P_J(\Omega)} \int_{\Omega} J(x - y)dy.$$

Then, taking

$$\hat{g}(x) = \frac{-\frac{P_J(\Omega)}{|\Omega|} + \int_{\mathbb{R}^N \setminus \Omega} J(x - y)dy}{\int_{\Omega} J(x - y)dy}, \quad x \in \Omega,$$

we have that $\tau(x) = 1$ for all $x \in \Omega$. Moreover, for $x \in \mathbb{R}^N \setminus \Omega$,

$$\tau(x) = -\frac{|\Omega|}{P_J(\Omega)} \int_{\Omega} J(x - y)dy \leq 0.$$

We claim now that

$$\frac{P_J(\Omega)}{|\Omega|} \tau \in \partial \mathcal{F}^J(0). \tag{5.9}$$

Take $w \in L^2(\mathbb{R}^N)$ with $\mathcal{TV}_J(w) < +\infty$. Since

$$w(x) = \int_0^{\infty} \chi_{E_t(w)}(x)dt - \int_{-\infty}^0 (1 - \chi_{E_t(w)})(x)dt,$$

we have

$$\int_{\mathbb{R}^N} \frac{P_J(\Omega)}{|\Omega|} \tau(x) w(x) dx = \frac{P_J(\Omega)}{|\Omega|} \int_{\mathbb{R}^N} \tau(x) \left(\int_{-\infty}^{\infty} \chi_{E_t(w)}(x) dt \right) dx$$

$$= \frac{P_J(\Omega)}{|\Omega|} \int_{-\infty}^{+\infty} \int_{\mathbb{R}^N} \tau(x) \chi_{E_t(w)}(x) dx dt.$$

Now, using that Ω is J-calibrable we have that

$$\frac{P_J(\Omega)}{|\Omega|} \int_{-\infty}^{+\infty} \int_{\mathbb{R}^N} \tau(x) \chi_{E_t(w)}(x) dx dt$$

$$= \frac{P_J(\Omega)}{|\Omega|} \int_{-\infty}^{+\infty} |E_t(w) \cap \Omega| dt + \frac{P_J(\Omega)}{|\Omega|} \int_{-\infty}^{+\infty} \int_{E_t(w) \setminus \Omega} \tau(x) dx dt$$

$$\leq \int_{-\infty}^{+\infty} P_J(E_t(w) \cap \Omega) dt + \frac{P_J(\Omega)}{|\Omega|} \int_{-\infty}^{+\infty} \int_{E_t(w) \setminus \Omega} \tau(x) dx dt.$$

By Proposition 1.2 and the coarea formula given in Theorem 2.4, we get

$$\int_{-\infty}^{+\infty} P_J(E_t(w) \cap \Omega) dt + \frac{P_J(\Omega)}{|\Omega|} \int_{-\infty}^{+\infty} \int_{E_t(w) \setminus \Omega} \tau(x) dx dt$$

$$= \int_{-\infty}^{+\infty} P_J(E_t(w) \cap \Omega) dt + \int_{-\infty}^{+\infty} P_J(E_t(w) \setminus \Omega) dt$$

$$- \int_{-\infty}^{+\infty} 2L_J(E_t(w) \setminus \Omega, E_t(w) \cap \Omega) dt - \int_{-\infty}^{+\infty} P_J(E_t(w) \setminus \Omega) dt$$

$$+ \int_{-\infty}^{+\infty} 2L_J(E_t(w) \setminus \Omega, E_t(w) \cap \Omega) dt + \frac{P_J(\Omega)}{|\Omega|} \int_{-\infty}^{+\infty} \int_{E_t(w) \setminus \Omega} \tau(x) dx dt$$

$$= \int_{-\infty}^{+\infty} P_J(E_t(w)) dt + I = \mathcal{TV}_J(w) + I,$$

with

$$I = - \int_{-\infty}^{+\infty} P_J(E_t(w) \setminus \Omega) dt + \int_{-\infty}^{+\infty} 2L_J(E_t(w) \setminus \Omega, E_t(w) \cap \Omega) dt$$

$$+ \frac{P_J(\Omega)}{|\Omega|} \int_{-\infty}^{+\infty} \int_{E_t(w) \setminus \Omega} \tau(x) dx dt.$$

Hence, if we prove that $I \leq 0$, we get

$$\int_{\mathbb{R}^N} \frac{P_J(\Omega)}{|\Omega|} \tau(x) w(x) dx \leq \mathcal{TV}_J(w). \tag{5.10}$$

Now, since

$$P_J(E_t(w) \setminus \Omega) = L_J(E_t(w) \setminus \Omega, \mathbb{R}^N \setminus (E_t(w) \setminus \Omega))$$

$$= L_J(E_t(w) \setminus \Omega, (E_t(w) \cap \Omega) \,\dot{\cup}\, (\mathbb{R}^N \setminus E_t(w))),$$

we have

$$I = -\int_{-\infty}^{+\infty} P_J(E_t(w) \setminus \Omega) dt + \int_{-\infty}^{+\infty} 2L_J(E_t(w) \setminus \Omega, E_t(w) \cap \Omega) dt$$

$$+ \frac{P_J(\Omega)}{|\Omega|} \int_{-\infty}^{+\infty} \int_{E_t(w) \setminus \Omega} \tau(x) dx dt$$

$$= -\int_{-\infty}^{+\infty} L_J(E_t(w) \setminus \Omega, \mathbb{R}^N \setminus E_t(w)) dt + \int_{-\infty}^{+\infty} L_J(E_t(w) \setminus \Omega, E_t(w) \cap \Omega) dt$$

$$+ \frac{P_J(\Omega)}{|\Omega|} \int_{-\infty}^{+\infty} \int_{E_t(w) \setminus \Omega} \tau(x) dx dt$$

$$= \int_{-\infty}^{+\infty} \left(\int_{E_t(w) \setminus \Omega} \left(\int_{\mathbb{R}^N \setminus E_t(w)} -J(x-y) dy \right. \right.$$

$$\left. \left. + \int_{E_t(w) \cap \Omega} J(x-y) dy + \frac{P_J(\Omega)}{|\Omega|} \tau(x) \right) dx \right) dt$$

$$= \int_{-\infty}^{+\infty} \left(\int_{E_t(w) \setminus \Omega} \left(\int_{\mathbb{R}^N \setminus E_t(w)} (-\tilde{g}(x,y) - 1) J(x-y) dy \right) dx \right) dt$$

$$+ \int_{-\infty}^{+\infty} \left(\int_{E_t(w) \setminus \Omega} \left(\int_{E_t(w) \cap \Omega} (-\tilde{g}(x,y) + 1) J(x-y) dy \right) dx \right) dt$$

$$- \int_{-\infty}^{+\infty} \left(\int_{E_t(w) \setminus \Omega} \left(\int_{E_t(w) \setminus \Omega} J(x-y) \tilde{g}(x,y) dy \right) dx \right) dt.$$

Now, the first integral is negative since $\tilde{g}(x,y) \geq -1$ for $x \in \mathbb{R}^N \setminus \Omega$, the second is zero since $\tilde{g}(x,y) = 1$ for $x \in \mathbb{R}^N \setminus \Omega$ and $y \in \Omega$ and the last integral is zero since $\tilde{g} = 0$ in $(\mathbb{R}^N \setminus \Omega) \times (\mathbb{R}^N \setminus \Omega)$. Therefore, $I \leq 0$, and consequently (5.10) holds.

Now, (5.10) implies that (5.9) is true. Then, by Theorem 4.5, we have

$$-\frac{P_J(\Omega)}{|\Omega|}\tau \in \Delta_1^J(0).$$

Thus, there exists $\mathbf{g} \in L^\infty(\mathbb{R}^N \times \mathbb{R}^N)$, $\mathbf{g}(x, y) = -\mathbf{g}(y, x)$ for almost all $(x, y) \in \mathbb{R}^N \times \mathbb{R}^N$, $\|\mathbf{g}\|_\infty \leq 1$, satisfying

$$\int_{\mathbb{R}^N} J(x - y)\mathbf{g}(x, y)\,dy = -\frac{P_J(\Omega)}{|\Omega|}\tau(x) \quad \text{a.e } x \in \mathbb{R}^N.$$

and

$$J(x - y)\mathbf{g}(x, y) \in J(x - y)\mathrm{sign}(0) \quad \text{a.e. } (x, y) \in \mathbb{R}^N \times \mathbb{R}^N.$$

Now, multiplying by χ_Ω and integrating by parts we get

$$P_J(\Omega) = \frac{P_J(\Omega)}{|\Omega|}\int_{\mathbb{R}^N}\tau(x)\chi_\Omega(x)dx$$

$$= -\int_{\mathbb{R}^N}\int_{\mathbb{R}^N} J(x - y)\mathbf{g}(x, y)\chi_\Omega(x)dxdy$$

$$= \frac{1}{2}\int_{\mathbb{R}^N}\int_{\mathbb{R}^N} J(x - y)\mathbf{g}(x, y)(\chi_\Omega(y) - \chi_\Omega(x))dxdy$$

$$\leq P_J(\Omega),$$

from where it follows that

$$J(x - y)\mathbf{g}(x, y) \in J(x - y)\mathrm{sign}(\chi_\Omega(y) - \chi_\Omega(x)) \quad \text{a.e. } (x, y) \in \mathbb{R}^N \times \mathbb{R}^N,$$

and consequently,

$$-\frac{P_J(\Omega)}{|\Omega|}\tau \in \Delta_1^J\chi_\Omega \quad \text{in } \mathbb{R}^N.$$

$\square$

Remark 5.13 It is not clear when Ω J-calibrable implies that

$$-\frac{P_J(\Omega)}{|\Omega|}\chi_\Omega \in \Delta_1^J\chi_\Omega \quad \text{in } \mathbb{R}^N.$$

We now give a result that says that J-calibrability is related to the nonlocal curvature function $H_{\partial E}^{J}$, which is the nonlocal version of one implication in the equivalence between (a) and (c) in Theorem 5.10.

Theorem 5.14 *Let $\Omega \subset \mathbb{R}^{N}$ be a bounded set and assume $\int_{\Omega} J(x - y)dy > 0$ for all $x \in \Omega$. Then,*

$$\Omega \text{ is } J\text{-calibrable} \Rightarrow \operatorname*{ess\,sup}_{x \in \Omega} H_{\partial \Omega}^{J}(x) \leq \frac{P_{J}(\Omega)}{|\Omega|}.$$

Proof By Theorem 5.12, there exists $\mathbf{g} \in L^{\infty}(\mathbb{R}^{N} \times \mathbb{R}^{N})$ antisymmetric with $\|\mathbf{g}\|_{\infty} \leq 1$ and a function τ, equal to 1 in Ω, such that

$$-\int_{\mathbb{R}^{N}} J(x - y)\mathbf{g}(x, y) \, dy = \frac{P_{J}(\Omega)}{|\Omega|}\tau(x) \quad \text{a.e } x \in \mathbb{R}^{N}, \tag{5.11}$$

and

$$J(x - y)\mathbf{g}(x, y) \in J(x - y)\operatorname{sign}(\mathcal{X}_{\Omega}(y) - \mathcal{X}_{\Omega}(x)) \quad \text{a.e. in } \mathbb{R}^{N} \times \mathbb{R}^{N}. \tag{5.12}$$

Then, by (5.11), (5.12) and since $\tau = 1$ in Ω, we have

$$H_{\partial \Omega}^{J}(x) = \int_{\mathbb{R}^{N}} J(x - y)(\mathcal{X}_{\mathbb{R}^{N} \setminus \Omega}(y) - \mathcal{X}_{\Omega}(y))dy$$

$$\leq -\int_{\mathbb{R}^{N}} J(x - y)\mathbf{g}(x, y) \, dy \leq \frac{P_{J}(\Omega)}{|\Omega|} \quad \text{for a.e. } x \in \Omega.$$

$\square$

In the next example we show that the reverse of Theorem 5.14 is not true in general.

Remark 5.15 Observe that

$$\operatorname*{ess\,sup}_{x \in \Omega} H_{\partial \Omega}^{J}(x) \leq \frac{P_{J}(\Omega)}{|\Omega|}$$

if and only if

$$\frac{1}{|\Omega|} \int_\Omega \int_\Omega J(x-y)dydx \leq 2 \operatorname*{ess\,inf}_{x\in\Omega} \int_\Omega J(x-y)dy. \tag{5.13}$$

Example 5.16 In general,

$$\operatorname*{ess\,sup}_{x\in\Omega} H^J_{\partial\Omega}(x) \leq \frac{P_J(\Omega)}{|\Omega|} \quad \text{does not imply } \Omega \text{ is } J\text{-calibrable.}$$

In fact, to provide an example where this happens, take $N = 3$ and the uniform $J = \frac{1}{|B_1|} \chi_{B_1}$. Let $0 < \epsilon < 1$, and $x_0 \in \mathbb{R}^N$ such that $\|x_0\| > 4$. Consider $\Omega_\epsilon = B_1 \cup B_{1+\epsilon}(x_0)$. For shortness, we write $B_{2,\epsilon} = B_{1+\epsilon}(x_0)$. Then, by Proposition 1.2 and (5.6), we have that

$$\frac{P_J(\Omega_\epsilon)}{|\Omega_\epsilon|} = \frac{P_J(B_1) + P_J(B_{2,\epsilon})}{|B_1| + |B_{2,\epsilon}|} > \frac{P_J(B_{2,\epsilon})}{|B_{2,\epsilon}|}.$$

Therefore, Ω_ϵ is not J-calibrable. Now, let us see that for ϵ small

$$\operatorname*{ess\,sup}_{x\in\Omega_\epsilon} H^J_{\partial\Omega_\epsilon}(x) \leq \frac{P_J(\Omega_\epsilon)}{|\Omega_\epsilon|}.$$

It is easy to see that

$$\operatorname*{ess\,sup}_{x\in\Omega_\epsilon} H^J_{\partial\Omega_\epsilon}(x) = \operatorname*{ess\,sup}_{x\in B_1} H^J_{\partial B_1}(x).$$

Let us see that there exists $\epsilon_0 < 1$ such that, for all $0 < \epsilon < \epsilon_0$:

$$\operatorname*{ess\,sup}_{x\in B_1} H^J_{\partial B_1}(x) \leq \frac{P_J(\Omega_\epsilon)}{|\Omega_\epsilon|};$$

which, by (5.13), is equivalent to prove that there exists $\epsilon_0 < 1$ such that, for all $0 < \epsilon < \epsilon_0$,

$$\frac{1}{|B_1| + |B_{2,\epsilon}|} \left(\int_{B_1} \int_{B_1} J(x-y)dydx + \int_{B_{2,\epsilon}} \int_{B_{2,\epsilon}} J(x-y)dydx \right)$$
$$\leq 2 \operatorname*{ess\,inf}_{x\in B_1} \int_{B_1} J(x-y)dy. \tag{5.14}$$

Now, if (5.14) is not true, then there exists $\epsilon_n \to 0$ such that

$$\frac{1}{|B_1| + |B_{2,\epsilon_n}|} \left(\int_{B_1} \int_{B_1} J(x-y) dy dx + \int_{B_{2,\epsilon_n}} \int_{B_{2,\epsilon_n}} J(x-y) dy dx \right) \tag{5.15}$$

$$\geq 2 \operatorname*{ess\,inf}_{x \in B_1} \int_{B_1} J(x-y) dy.$$

Hence, passing to the limit in (5.15), we get

$$\frac{1}{|B_1|} \int_{B_1} \int_{B_1} J(x-y) dy dx \geq 2 \operatorname*{ess\,inf}_{x \in B_1} \int_{B_1} J(x-y) dy.$$

But, as we see below, the inequality in (5.13) is strict and hence the above inequality gives a contradiction.

Let us show that

$$\frac{1}{|B_1|} \int_{B_1} \int_{B_1} J(x-y) dy dx < 2 \operatorname*{ess\,inf}_{x \in B_1} \int_{B_1} J(x-y) dy.$$

Since $J = \frac{1}{|B_1|} \chi_{B_1}$, it holds that

$$\int_{B_1} J(x-y) dy = \frac{1}{|B_1|} \int_{B_1} \chi_{B_1}(x-y) dy = \frac{1}{|B_1|} |B_1 \cap B_1(x)|,$$

and then the previous inequality turns out to be

$$\frac{1}{|B_1|} \int_{B_1} |B_1 \cap B_1(x)| dx < 2 \operatorname*{ess\,inf}_{x \in B_1} |B_1 \cap B_1(x)|. \tag{5.16}$$

Now, since

$$|B_1 \cap B_1(x)| = \frac{\pi}{12}(|x| + 4)(2 - |x|)^2,$$

we have that

$$\frac{1}{|B_1|} \int_{B_1} |B_1 \cap B_1(x)| dx = \frac{1}{16} \int_{B_1} (|x| + 4)(2 - |x|)^2 dx$$

$$= \frac{\pi}{8} \int_0^1 r(r + 4)(2 - r)^2 dr$$

$$= \frac{21\pi}{40}.$$

On the other hand,

$$2\operatorname*{ess\,inf}_{x\in B_1}|B_1 \cap B_1(x)| = 2|B_1 \cap B_1(0,0,1)| = \frac{5\pi}{6},$$

and we conclude that (5.16) holds.

Note that in this example Ω is not connected. However, this fact is not relevant. We have that the nonlocal perimeter and the nonlocal curvature are continuous with respect to the set in terms of convergence in measure (if $E_n \to E$ in the sense that $|E_n \triangle E| \to 0$, then $P_J(E_n) \to P_J(E)$ and $H^J_{\partial E_n}(x) \to H^J_{\partial E}(x)$). Then, we only have to connect the two balls with a thin bridge to obtain an example of a connected domain such that

$$\operatorname*{ess\,sup}_{x\in\Omega} H^J_{\partial\Omega}(x) \leq \frac{P_J(\Omega)}{|\Omega|} \quad \text{but } \Omega \text{ is not } J\text{-calibrable.}$$

Example 5.17 Let $\Omega \subset \mathbb{R}^N$ be a bounded set and assume $\int_\Omega J(x-y)dy > 0$ for all $x \in \Omega$. One can ask if

$$\sup_{x\in\partial\Omega} H^J_{\partial\Omega}(x) \leq \frac{P_J(\Omega)}{|\Omega|}. \tag{5.17}$$

implies J-calibrability. The following example shows that, in general, this is not the case. For $N = 1$, and for $J(z) = \frac{1}{2}\chi_{[-1,1]}(z)$ and

$$\Omega_\epsilon = (]-3,3[\setminus[-1,1])\cup]-\epsilon,\epsilon[, \quad 0 < \epsilon < 1/2,$$

one can check that

$$\sup_{x\in\partial\Omega_\epsilon} H^J_{\partial\Omega_\epsilon}(x) = (1-3\epsilon)^+$$

$$< H^J_{\partial\Omega_\epsilon}(0) = 1 - 2\epsilon$$

$$= \sup_{x\in\Omega} H^J_{\partial\Omega_\epsilon}(x),$$

and that

$$\frac{P_J(\Omega_\epsilon)}{|\Omega_\epsilon|} = \frac{1+2\epsilon-3\epsilon^2}{4+2\epsilon}.$$

Then, for $\epsilon < \sqrt{19}-4$,

$$1 - 2\epsilon > \frac{P_J(\Omega_\epsilon)}{|\Omega_\epsilon|}$$

and consequently Ω is not calibrable. Now, for

$$0.236 \simeq \sqrt{5} - 2 \le \epsilon < \sqrt{19} - 4 \simeq 0.359$$

it holds that

$$\sup_{x \in \partial \Omega_\epsilon} H^J_{\partial \Omega_\epsilon}(x) \le \frac{P_J(\Omega_\epsilon)}{|\Omega_\epsilon|},$$

and consequently condition (5.17) is not enough for J-calibrability.

Example 5.18 Let J be a nonnegative radially nonincreasing function. It is easy to see that a large cube $] - L, L[^N$ is not J-calibrable; this is also true in the local case for any cube. Nevertheless, in the local case, the stadium given by the convex hull of two circles ($N = 2$):

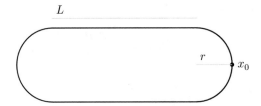

$$\Omega_r^L = \mathrm{co}\Big(B_r(-\frac{L}{2}, 0) \cup B_r(\frac{L}{2}, 0)\Big),$$

with $L, r > 0$, is calibrable. Indeed,

$$\operatorname{ess\,sup}_{x \in \partial \Omega_r^L} H_{\partial \Omega_r^L}(x) = \frac{1}{r} < \frac{2\pi r + 2L}{\pi r^2 + 2rL} = \frac{\mathrm{Per}(\Omega)}{|\Omega|},$$

and Theorem 5.10 gives the result. But, Ω_r^L is not J-calibrable for the uniform $J = \frac{1}{|B_1(0,0)|} \chi_{B_1(0,0)}$, $r > 1$ and $L > 2$ large enough. Moreover, we have that this non-J-calibrable set does not contain a J-Cheeger set.

Let us prove these two facts. For the first one, by Theorem 5.14 and Remark 5.15, it is enough to show that, for L large enough,

$$\frac{1}{|\Omega_r^L|} \int_{\Omega_r^L} \int_{\Omega_r^L} J(x - y) dy dx > 2 \operatorname*{ess\,inf}_{x \in \Omega_r^L} \int_{\Omega_r^L} J(x - y) dy.$$

Now, this condition reads as follows:

$$\frac{1}{|\Omega_r^L|} \int_{\Omega_r^L} |\Omega_r^L \cap B_1(x)| dx > 2 \left| B_1(\frac{L}{2} + r, 0) \cap B_r(\frac{L}{2}, 0) \right|, \qquad (5.18)$$

since

$$\operatorname*{ess\,inf}_{x \in \Omega_r^L} \int_{\Omega_r^L} J(x - y) dy = \int_{\Omega_r^L} J(x_0 - y) dy$$

for $x_0 = (\frac{L}{2} + r, 0)$. Let us call $a(r) = 2 \left| B_1(\frac{L}{2} + r, 0) \cap B_r(\frac{L}{2}, 0) \right|$. Now, if $C_r^L =] - \frac{L}{2}, \frac{L}{2}[\times]0, r - 1[$ and $D_r^L =]1 - \frac{L}{2}, \frac{L}{2} - 1[\times]r - 1, r[$, we have

$$\frac{1}{|\Omega_r^L|} \int_{\Omega_r^L} |\Omega_r^L \cap B_1(x)| dx \geq \frac{2}{\pi r^2 + 2rL} \left(\pi L(r - 1) + \int_{D_r^L} |\Omega_r^L \cap B_1(x)| dx \right)$$

$$\geq \frac{2}{\pi r^2 + 2rL} \left(\pi L(r - 1) + (L - 2)(\pi - \frac{2}{3}) \right),$$

and this last quantity is larger than $a(r)$ if

$$L > \frac{\frac{1}{2}a(r)\pi r^2 + \pi - \frac{2}{3}}{r\left(\pi - \frac{2}{3r} - a(r)\right)},$$

since $\pi - \frac{2}{3r} > a(r)$. That is, we get that (5.18) holds true for L large enough. Consequently, Ω_r^L is not J-calibrable.

Let us now see that Ω_r^L does not contain a J-Cheeger set. Arguing by contradiction, assume that there exists E a J-Cheeger set of Ω_r^L. Then, since E is J-calibrable and having in mind the calculation made above, we have

$$a(r) = \frac{2}{\pi} \left| B_1(\frac{L}{2} + r, 0) \cap B_r(\frac{L}{2}, 0) \right|$$

$$< \frac{1}{|\Omega_r^L|} \int_{\Omega_r^L} \int_{\Omega_r^L} J(x - y) dy dx$$

$$< \frac{1}{|E|} \int_E \int_E J(x - y) dy dx$$

$$\leq 2 \inf_{x \in E} \int_E J(x - y) dy$$

$$= \frac{2}{\pi} \inf_{x \in E} |B_1(x) \cap E|.$$

On the other hand, consider a ball B_s such that $|B_s| = |E|$. By the Isoperimetric Inequality and since B_s is J-calibrable, we have

$$a(r) < \frac{1}{|E|} \int_E \int_E J(x - y) dy dx$$

$$\leq \frac{1}{|B_s|} \int_{B_s} \int_{B_s} J(x - y) dy dx$$

$$\leq a(s),$$

from where it follows that $s > r$, and consequently $E \not\subset B_r(-\frac{L}{2}, 0)$. Hence, if we define

$$l^+ = \sup \left\{ l \in \left[-\frac{L}{2}, \frac{L}{2} \right] \ : \ |B_r(l, 0) \cap E| > 0 \right\},$$

we have $l^+ > -\frac{L}{2}$. Therefore, for $-\frac{L}{2} < l < l^+$, we have

$$a(r) < \frac{2}{\pi} \operatorname*{ess\,inf}_{x \in E} |B_1(x) \cap E|$$

$$\leq \frac{2}{\pi} \operatorname{ess\,inf} \left\{ |B_1(x) \cap E| \ : \ x \in E \cap (B_r(l^+, 0) \setminus B_r(l, 0)) \right\}$$

$$\leq a(r) + o(l^+ - l).$$

Then, letting $l \to l^+$ we arrive to a contradiction.

5.3 Rescaling

In this section, we relate local and nonlocal Cheeger constants and local and nonlocal calibrable sets under rescaling.

Proposition 5.19 *Let $N \geq 2$. Assume J satisfies (1.19). Let Ω be an open bounded set of $\mathbb{R}^N$, then*

$$\lim_{\epsilon \downarrow 0} C_{J_\epsilon} h_1^{J_\epsilon}(\Omega) = h_1(\Omega).$$

Proof Given $\delta > 0$, there exists $E_\delta \subset \Omega$ such that

$$h_1(\Omega) + \delta \geq \frac{\operatorname{Per}(E_\delta)}{|E_\delta|}.$$

Then, by Theorem 1.11, we have

$$h_1(\Omega) + \delta \geq \lim_{\epsilon \downarrow 0} \frac{C_J}{\epsilon} \frac{P_{J_\epsilon}(E_\delta)}{|E_\delta|} \geq \limsup_{\epsilon \downarrow 0} \frac{C_J}{\epsilon} h_1^{J_\epsilon}(\Omega).$$

By the arbitrariness of δ, we get

$$\limsup_{\epsilon \downarrow 0} \frac{C_J}{\epsilon} h_1^{J_\epsilon}(\Omega) \leq h_1(\Omega).$$

Let us now suppose that

$$\liminf_{\epsilon \downarrow 0} \frac{C_J}{\epsilon} h_1^{J_\epsilon}(\Omega) < h_1(\Omega). \tag{5.19}$$

By (5.4), given $\epsilon > 0$, there exists $u_\epsilon \in BV_{J_\epsilon}(\Omega)$, $u_\epsilon = 0$ in $\mathbb{R}^N \setminus \Omega$, $\|u_\epsilon\| = 1$, such that

$$h_1^{J_\epsilon}(\Omega) \leq \mathcal{TV}_{J_\epsilon}(u_\epsilon) \leq h_1^{J_\epsilon}(\Omega) + \epsilon^2.$$

Then, by (5.19),

$$\liminf_{\epsilon \downarrow 0} \frac{C_J}{\epsilon} \mathcal{TV}_{J_\epsilon}(u_\epsilon) < h_1(\Omega).$$

Therefore, there exists a sequence ϵ_n decreasing to 0 such that

$$\frac{C_J}{\epsilon_n} \mathcal{TV}_{J_{\epsilon_n}}(u_{\epsilon_n}) < h_1(\Omega).$$

Then, for a large ball B containing Ω,

$$\frac{C_J}{\epsilon_n} \mathcal{TV}_{J_{\epsilon_n}}(u_{\epsilon_n}) = \frac{1}{K_{1,N}} \int_B \int_B \frac{|u_{\epsilon_n}(x) - u_{\epsilon_n}(y)|}{|x - y|} \rho_{\epsilon_n}(x - y) dx dy < h_1(\Omega),$$

where

$$\rho_\epsilon(z) = \frac{1}{2} C_J K_{1,N} \frac{|z|}{\epsilon} J_\epsilon(z).$$

Consequently, by Theorem 1.9 (observe that ρ_ϵ satisfies (1.26)), we have that there exists a subsequence of ϵ_n, denoted equal, such that

$$u_{\epsilon_n} \to u \quad \text{in } L^1(B),$$

$u \in BV(B)$ and, since moreover $u = u\chi_\Omega$,

$$\int_B |Du| = \int_\Omega |Du| + \int_{\partial\Omega} |u| d\mathcal{H}^{N-1} < h_1(\Omega).$$

But, we also get that $\|u\|_{L^1(\Omega)} = 1$ and consequently, from the above inequality, we get $\lambda_1(\Omega) < h_1(\Omega)$, which is a contradiction. Therefore, what we supposed in (5.19) is false and then

$$\liminf_{\epsilon \downarrow 0} \frac{C_J}{\epsilon} h_1^{J_\epsilon}(\Omega) \geq h_1(\Omega),$$

and the proof concludes.   □

Note that there are sets which are J_ϵ-calibrable for every $\epsilon > 0$ small, for example a fixed ball. Our next result says that such sets are calibrable.

Corollary 5.20 *Let $N \geq 2$. Assume J satisfies (1.19). Let Ω be an open bounded set of $\mathbb{R}^N$. If Ω is J_{ϵ_n}-calibrable for a sequence $\epsilon_n \to 0$ as $n \to +\infty$, then Ω is calibrable.*

Proof Since Ω is J_{ϵ_n}-calibrable, we have

$$\frac{C_J}{\epsilon_n} h_1^{J_{\epsilon_n}}(\Omega) = \frac{C_J}{\epsilon_n} \frac{P_{J_{\epsilon_n}}(\Omega)}{|\Omega|}.$$

Hence, by Theorem 1.11,

$$\frac{C_J}{\epsilon_n} h_1^{J_{\epsilon_n}}(\Omega) \to \frac{\text{Per}(\Omega)}{|\Omega|} \quad \text{as } n \to +\infty.$$

Then, by Proposition 5.19, we conclude that

$$\frac{\text{Per}(\Omega)}{|\Omega|} = h_1(\Omega),$$

and consequently Ω is calibrable.   □

Nonlocal Heat Content

6

6.1 The Classical Heat Content

The *heat content* of a Borel measurable set $D \subset \mathbb{R}^N$ at time t is defined by M. van der Berg in [69] (see also [70]) as:

$$\mathbb{H}_D(t) = \int_D T(t) \chi_D(x) dx,$$

with $(T(t))_{t \geq 0}$ being the heat semigroup in $L^2(\mathbb{R}^N)$. Therefore, the heat content represents the amount of heat in D at time t if in D the initial temperature is 1 and in $\mathbb{R}^N \setminus D$ the initial temperature is 0.

The following characterization for the perimeter of a set $D \subset \mathbb{R}^N$ with finite perimeter was given in [61, Theorem 3.3]:

$$\lim_{t \to 0^+} \sqrt{\frac{\pi}{t}} \int_{\mathbb{R}^N \setminus D} T(t) \chi_D(x) dx = \mathrm{Per}(D), \qquad (6.1)$$

As a consequence, the following result was presented in [69] for the heat content:

For an open subset D in $\mathbb{R}^N$ with finite Lebesgue measure and finite perimeter, there holds

$$\mathbb{H}_D(t) = |D| - \sqrt{\frac{t}{\pi}} \mathrm{Per}(D) + o(\sqrt{t}) \quad \text{as } t \downarrow 0. \qquad (6.2)$$

In [3] and [2], the concept of heat content has been extended to more general diffusion processes (see also [35]). More precisely, for $0 < \alpha \leq 2$, let $p_t^{(\alpha)} : \mathbb{R}^N \to [0, \infty)$ be the

© Springer Nature Switzerland AG 2019
J. M. Mazón et al., *Nonlocal Perimeter, Curvature and Minimal Surfaces for Measurable Sets*, Frontiers in Mathematics,
https://doi.org/10.1007/978-3-030-06243-9_6

probability density such that its Fourier transform verifies

$$\widehat{p_t^{(\alpha)}}(x) = e^{-t|x|^\alpha}.$$

If one considers

$$T_t^{(\alpha)}(f)(x) = \int_{\mathbb{R}^N} f(y) p_t^{(\alpha)}(x - y)dy,$$

then $u(x, t) = T_t^{(\alpha)}(f)(x)$ is the unique weak solution of the initial valued problem:

$$\begin{cases} u_t(x, t) = -(-\Delta)^{\frac{\alpha}{2}} u(x, t), \ (x, t) \in \mathbb{R}^N \times [0, \infty), \\ \\ u(x, 0) = f(x) \qquad\qquad\quad x \in \mathbb{R}^N. \end{cases}$$

And, in this context, the *heat content* of a Borel measurable set $D \subset \mathbb{R}^N$ at time t is defined as:

$$\mathbb{H}_D^{(\alpha)}(t) = \int_D T_t^{(\alpha)} \chi_D(x)dx.$$

Note that for $\alpha = 2$,

$$p_t^{(2)} = (4\pi t)^{-\frac{N}{2}} \exp\left(-\frac{|x|^2}{4t}\right),$$

is the Gaussian kernel and consequently $\mathbb{H}_D^{(2)}(t) = \mathbb{H}_D(t)$; while, for $\alpha = 1$,

$$p_t^{(1)}(x) = \frac{\Gamma\left(\frac{N+1}{2}\right)}{\pi^{\frac{N+1}{2}}} \frac{t}{(t + |x|)^{\frac{N+1}{2}}}$$

is the Poisson heat kernel. In [3] and [2], it is proved that, for bounded sets D in $\mathbb{R}^N$ of finite perimeter, $N \geq 2$,

$$\lim_{t \to 0^+} \frac{1}{t}\left(|D| - \mathbb{H}_D^{(\alpha)}(t)\right) = A_{\alpha,N} \mathrm{Per}_\alpha(D)$$

for $0 < \alpha < 1$, where $A_{\alpha,N}$ is a determined positive constant,

$$\lim_{t \to 0^+} \frac{1}{t^{1/\alpha}}\left(|D| - \mathbb{H}_D^{(\alpha)}(t)\right) = \frac{1}{\pi}\Gamma(1 - 1/\alpha)\mathrm{Per}(D)$$

for $1 < \alpha < 2$ and

$$\lim_{t \to 0^+} \frac{1}{t \ln(1/t)} \left(|D| - \mathbb{H}_D^{(1)}(t) \right) = \frac{1}{\pi} \mathrm{Per}(D)$$

for D smooth.

6.2   The Heat Content for Nonlocal Diffusion with Nonsingular Kernels

6.2.1   Definition and a Characterization

Assume J to be continuous with $J(0) > 0$ and compactly supported. The following nonlocal Cauchy problem has been studied in [12]:

$$\begin{cases} u_t(x, t) = \displaystyle\int_{\mathbb{R}^N} J(x - y)(u(y, t) - u(x, t)) dy, \ (x, t) \in \mathbb{R}^N \times [0, +\infty), \\ u(x, 0) = u_0(x), \hspace{5.5cm} x \in \mathbb{R}^N. \end{cases} \tag{6.3}$$

This equation appears naturally from the following considerations: if $u(x, t)$ is thought of as a density at a point x at time t, and $J(x - y)$ is thought of as the probability distribution of jumping from location y to location x, then

$$\int_{\mathbb{R}^N} J(y - x)u(y, t)\, dy = (J * u)(x, t)$$

is the rate at which individuals are arriving at position x from all other places, and

$$-u(x, t) = -\int_{\mathbb{R}^N} J(y - x)u(x, t)\, dy$$

is the rate at which they are leaving location x to travel to all other sites. This consideration, in the absence of external or internal sources, leads immediately to the fact that the density u satisfies Eq. (6.3).

In [12], it is defined a solution of problem (6.3) in the time interval $[0, T]$ as a function $u \in W^{1,1}(0, T; L^2(\mathbb{R}^N))$ which satisfies $u(x, 0) = u_0$ and

$$u_t(x, t) = \int_{\mathbb{R}^N} J(x - y)(u(y, t) - u(x, t))\, dy \quad \text{a.e. in } (x, t) \in \mathbb{R}^N \times (0, T).$$

A simple integration of the equation in (6.3) in space gives that the total mass is preserved, that is:

$$\int_{\mathbb{R}^N} u(x,t)dx = \int_{\mathbb{R}^N} u_0(x)dx \qquad \forall t \geq 0. \tag{6.4}$$

Our aim in this chapter is to study the heat content associated with the above nonlocal diffusion process and to relate it with the (local) heat content.

We give the following definition.

Definition 6.1 Given a Lebesgue measurable set $D \subset \mathbb{R}^N$ with finite measure, we define the *J-heat content* of D in $\mathbb{R}^N$ at time t by:

$$\mathbb{H}_D^J(t) = \int_D u(x,t)dx,$$

with u being the solution of (6.3) with the datum $u_0 = \chi_D$.

Note that, from (6.4), we have

$$\mathbb{H}_D^J(0) = |D|,$$

and that no further regularity is required for D besides having finite measure.

We have the following interpretation of the J-heat content $\mathbb{H}_D^J(t)$: since $u(x,t)$ represents the density of a population at a point $x \in \mathbb{R}^N$ at time t with initial condition $u(x,0) = \chi_D(x)$, then the J-heat content of D at time t represents the size of the population that remains inside D at that time when in D the initial density of the population is 1 and in $\mathbb{R}^N \setminus D$ the initial density of population is 0.

Our first result relates the J-heat content with the L^2-norm of the solution of (6.3). Let B be the operator defined in $L^2(\mathbb{R}^N)$ as:

$$B(u) = v \iff v(x) = -\Delta_J u(x) \quad \forall x \in \Omega,$$

where

$$\Delta_J u(x) = \int_{\mathbb{R}^N} J(x-y)(u(y) - u(x))dy.$$

The domain of B is $L^2(\mathbb{R}^N)$ and, in [12], it is showed that B is m-completely accretive in $L^2(\Omega)$. Then, this operator B generates a C_0-semigroup $(T_J(t))_{t\geq 0}$ in $L^2(\Omega)$ which solves the Cauchy problem (6.3). With this operator, we can describe the J-heat content of a finite measurable set $D \subset \mathbb{R}^N$ as:

$$\mathbb{H}_D^J(t) = \int_D T_J(t) \chi_D(x) dx,$$

and we have the following result.

Proposition 6.2 *Given a Lebesgue measurable set $D \subset \mathbb{R}^N$ with finite measure, we have*

$$\mathbb{H}_D^J(t) = \left\| T_J\left(\frac{t}{2}\right) \chi_D \right\|_{L^2}^2.$$

Proof It is enough to prove that the operators $T_J(t)$ are self-adjoint, since then

$$\left\| T_J\left(\frac{t}{2}\right) \chi_D \right\|_{L^2}^2 = \left\langle T_J\left(\frac{t}{2}\right) \chi_D, T_J\left(\frac{t}{2}\right) \chi_D \right\rangle$$

$$= \langle T_J(t) \chi_D, \chi_D \rangle$$

$$= \mathbb{H}_D^J(t).$$

So, let us see that $T_J(t)$ is self-adjoint.

Let $\mathcal{F} : L^2(\mathbb{R}^N) \to L^2(\mathbb{R}^N)$ be the Fourier–Plancherel transform. We will also write $\hat{f} = \mathcal{F}(f)$.

If $u(t)(x) = u(x, t) := T_J(t) f(x)$, since

$$u_t(t) = J * u(t) - u(t), \tag{6.5}$$

applying the Fourier–Plancherel transform, we have

$$\hat{u}_t(\xi, t) = (\hat{J}(\xi) - 1)\hat{u}(\xi, t), \tag{6.6}$$

from which it follows that

$$\hat{u}(\xi, t) = e^{(\hat{J}(\xi)-1)t} \widehat{f}(\xi).$$

Therefore, given $f, g \in L^2(\mathbb{R}^N)$, we have

$$\langle T_J(t)f, g \rangle = \langle \mathcal{F}(T_J(t)f), \mathcal{F}(g) \rangle$$

$$= \langle e^{(\hat{J}(\xi)-1)t}\widehat{f}, \widehat{g} \rangle$$

$$= \langle \widehat{f}, e^{(\hat{J}(\xi)-1)t}\widehat{g} \rangle$$

$$= \langle f, T_J(t)g \rangle,$$

as we wanted to show.   □

6.2.2   The Asymptotic Expansion of the J-heat Content

In this section, we give an asymptotic expansion of the J-heat content, from which a result similar to (6.2) follows. This expansion, obtained from a classical Taylor's expansion, is given for any time and for any order and with the main fact that all the terms involved in the expansion can be expressed using nonlocal perimeters of the set D with different kernels.

We will use the following notation:

$$J = (J*)^1,$$

$$J * J = (J*)^2,$$

and for the convolution of n kernels:

$$J * J * \ldots * J = (J*)^n.$$

Observe that, for all n, the kernel $(J*)^n$ satisfies the same structural conditions than J, in particular

$$\int_{\mathbb{R}^N} (J*)^n(x)dx = 1.$$

For these kernels, we can define, as in (1.9), the $(J*)^k$-nonlocal perimeter of a measurable set D:

$$P_{(J*)^k}(D) = \int_D \int_{\mathbb{R}^N \setminus D} (J*)^k(x - y)\, dy\, dx,$$

and the $(J*)^k$-nonlocal curvature of D at x:

$$H_{\partial D}^{(J*)}(x) = -\int_{\mathbb{R}^N} (J*)(x-y)(X_D(y) - X_{\mathbb{R}^N \backslash D}(y))dy.$$

We will also use the following relations (see the previous chapters). For a finite measurable set D,

$$|D| = \int_D \int_D (J*)^k(x-y)dydx + P_{(J*)^k}(D), \tag{6.7}$$

$$\mathcal{H}_{\partial D}^{(J*)^k}(x) = 1 - 2\int_D (J*)^k(x-y)dy \tag{6.8}$$

and hence

$$\int_D \mathcal{H}_{\partial D}^{(J*)^k}(x)\, dx = \int_D \left(1 - 2\int_D (J*)^k(x-y)dy\right)dx$$

$$= |D| - 2\int_D \int_D (J*)^k(x-y)\, dy\, dx.$$

that is,

$$\int_D \mathcal{H}_{\partial D}^{(J*)^k}(x)dx = 2P_{(J*)^k}(D) - |D|. \tag{6.9}$$

Proposition 6.3 *Let D be a finite measurable subset of $\mathbb{R}^N$ and let $u(\cdot, t) = T_J(t)X_D$. Then, the following hold:*

1. The Fourier transform of u verifies

$$\hat{u}(\xi, t) = \sum_{n=0}^{L}\left(\widehat{X_D}(\xi) + \sum_{k=1}^{n}\binom{n}{k}(-1)^{n-k}\hat{J}(\xi)^k\widehat{X_D}(\xi)\right)\frac{t^n}{n!}$$

$$+(\hat{J}(\xi) - 1)^{L+1}\hat{u}(\xi, s)\frac{t^{L+1}}{(L+1)!}, \tag{6.10}$$

for $0 < s < t$.
2. The following expansion holds for u:

$$u(x, t) = e^{-t}\sum_{n=0}^{\infty}\int_D (J*)^n(x-y)dy\frac{t^n}{n!}, \tag{6.11}$$

where

$$\int_D (J*)^0(x - y)dy = X_D(x).$$

Proof From (6.6), the Fourier transform of u verifies the evolution problem:

$$\hat{u}_t(\xi, t) = (\hat{J}(\xi) - 1)\hat{u}(\xi, t),$$

with the initial condition $\hat{u}(\xi, 0) = \widehat{X_D}(\xi)$. Hence,

$$\hat{u}(\xi, t) = e^{(\hat{J}(\xi)-1)t}\widehat{X_D}(\xi). \tag{6.12}$$

Using Taylor expansion of the exponential, we deduce

$$\hat{u}(\xi, t) = \sum_{n=0}^{L}(\hat{J}(\xi) - 1)^n\widehat{X_D}(\xi)\frac{t^n}{n!} + (\hat{J}(\xi) - 1)^{L+1}e^{(\hat{J}(\xi)-1)s}\widehat{X_D}(\xi)\frac{t^{L+1}}{(L+1)!}$$

$$= \sum_{n=0}^{L}\left(\widehat{X_D}(\xi) + \sum_{k=1}^{n}\binom{n}{k}(-1)^{n-k}\hat{J}(\xi)^k\widehat{X_D}(\xi)\right)\frac{t^n}{n!}$$

$$+ (\hat{J}(\xi) - 1)^{L+1}e^{(\hat{J}(\xi)-1)s}\widehat{X_D}(\xi)\frac{t^{L+1}}{(L+1)!}$$

$$= \sum_{n=0}^{L}\left(\widehat{X_D}(\xi) + \sum_{k=1}^{n}\binom{n}{k}(-1)^{n-k}\hat{J}(\xi)^k\widehat{X_D}(\xi)\right)\frac{t^n}{n!}$$

$$+ (\hat{J}(\xi) - 1)^{L+1}\hat{u}(\xi, s)\frac{t^{L+1}}{(L+1)!},$$

for $0 < s < t$.

From (6.12), we also have

$$\hat{u}(\xi, t) = e^{-t}\widehat{X_D}(\xi)e^{t\hat{J}(\xi)} = e^{-t}\sum_{n=0}^{\infty}(\hat{J}(\xi))^n\widehat{X_D}(\xi)\frac{t^n}{n!}.$$

Taking here the inverse Fourier transform, we get that

$$u(x, t) = e^{-t}\sum_{n=0}^{\infty}\mathcal{F}^{-1}\left((\hat{J}(\xi))^n\widehat{X_D}(\xi)\right)(x, t)\frac{t^n}{n!}$$

$$= e^{-t}\sum_{n=0}^{\infty}\int_D (J*)^n(x - y)dy\frac{t^n}{n!},$$

where

$$\int_D (J*)^0(x-y)dy = \chi_D(x).$$

$\square$

The main result of this section is the following.

Theorem 6.4 *Let D be a finite measurable set. Then,*

$$\mathbb{H}_D^J(t) = |D| - \sum_{n=1}^{+\infty} \left(\sum_{k=1}^n \binom{n}{k} (-1)^{n-k} P_{(J*)^k}(D) \right) \frac{t^n}{n!} \quad \forall t > 0. \tag{6.13}$$

Moreover, for $L \geq 1$ we have

$$\left| \mathbb{H}_D^J(t) - |D| + \sum_{n=1}^L \left(\sum_{k=1}^n \binom{n}{k} (-1)^{n-k} P_{(J*)^k}(D) \right) \frac{t^n}{n!} \right| \leq \frac{|D| 2^L}{(L+1)!} t^{L+1} \quad \forall t > 0. \tag{6.14}$$

We also have the following expression:

$$\mathbb{H}_D^J(t) = \sum_{n=0}^{+\infty} \left(\int_D \int_D (J*)^n(x-y)dydx \right) \frac{e^{-t}t^n}{n!} \quad \forall t > 0, \tag{6.15}$$

where

$$\int_D \int_D (J*)^0(x-y)dydx = |D|.$$

Proof Taking the inverse Fourier–Plancherel transform in (6.10) and integrating over D, we get

$$\mathbb{H}_D^J(t) = \sum_{n=0}^L \left((-1)^n |D| + \sum_{k=1}^n \binom{n}{k} (-1)^{n-k} \int_D \int_D (J*)^k(x-y)dydx \right) \frac{t^n}{n!}$$

$$+ \int_D \mathcal{F}^{-1} \left((\hat{J}(\xi) - 1)^{L+1} \hat{u}(\xi, s) \right)(x)dx \frac{t^{L+1}}{(L+1)!}$$

for $n \geq 1$. Now, for $n \geq 1$, using (6.7) yields

$$(-1)^n|D| + \sum_{k=1}^{n} \binom{n}{k}(-1)^{n-k} \int_D \int_D (J*)^k(x-y)dydx$$

$$= (-1)^n|D| + \sum_{k=1}^{n} \binom{n}{k}(-1)^{n-k}\left(|D| - P_{(J*)^k}(D)\right)$$

$$= -\sum_{k=1}^{n} \binom{n}{k}(-1)^{n-k} P_{(J*)^k}(D).$$

Also,

$$\int_D \mathcal{F}^{-1}\left((\hat{J}(\xi) - 1)^{L+1}\hat{u}(\xi, s)\right)(x)dx$$

$$= (-1)^{L+1}|D| + \sum_{k=1}^{L+1} \binom{L+1}{k}(-1)^{L+1-k} \int_D (J*)^k * u(x, s)dx.$$

(6.16)

Using the fact that for this problem, thanks to the maximum principle, we have

$$0 \leq u \leq 1,$$

we get

$$0 \leq \int_D ((J*)^k * u)(x, t)dx \leq |D| \qquad \forall k \in \mathbb{N}.$$

Hence, from (6.16) we have that

$$\left|\int_D \mathcal{F}^{-1}\left((\hat{J}(\xi) - 1)^{L+1}\hat{u}(\xi, s)\right)(x)dx\right| \leq \frac{1}{2}\sum_{k=0}^{L+1}\binom{L+1}{k}|D| = 2^L|D|.$$

From which (6.13) and (6.14) follow.

For the second statement in the thesis, let us integrate over D in (6.11), then we get

$$\mathbb{H}_D^J(t) = e^{-t}\sum_{n=0}^{\infty}\int_D\int_D (J*)^n(x-y)dydx\frac{t^n}{n!},$$

and thus (6.15) is proved.   □

We can also proceed in the following way. Integrating in (6.3) over D, we get

$$(\mathbb{H}_D^J)'(t) = \int_D \int_{\mathbb{R}^N} J(x-y)u(y,t)dydx - \int_D u(x,t)dx$$

$$= \int_D \int_{\mathbb{R}^N} J(x-y)u(y,t)dydx - \mathbb{H}_D^J(t),$$

and hence,

$$\mathbb{H}_D^J(t) + (\mathbb{H}_D^J)'(t) = \int_D \int_{\mathbb{R}^N} J(x-y)u(y,t)dydx. \tag{6.17}$$

Taking the time derivative in (6.17) and using again (6.17), we get

$$(\mathbb{H}_D^J)'(t) + (\mathbb{H}_D^J)''(t) = \int_D \int_{\mathbb{R}^N} J(x-y)u_t(y,t)dydx$$

$$= \int_D \int_{\mathbb{R}^N} J(x-y)\left(\int_{\mathbb{R}^N}(J(y-z)u(z,t)-u(y,t))dx\right)dydx$$

$$= \int_D \int_{\mathbb{R}^N} \int_{\mathbb{R}^N} J(x-y)J(y-z)u(z,t)dzdydx$$

$$- \int_D \int_{\mathbb{R}^N} J(x-y)u(y,t)dydx$$

$$= \int_D ((J*)^2 * u)(x,t)dx - \mathbb{H}_D^J(t) - (\mathbb{H}_D^J)'(t),$$

hence,

$$\mathbb{H}_D^J(t) + 2(\mathbb{H}_D^J)'(t) + (\mathbb{H}_D^J)''(t) = \int_D ((J*)^2 * u)(x,t)dx.$$

By induction, it is easy to see that

$$\sum_{k=0}^n \binom{n}{k}(\mathbb{H}_D^J)^{(k)}(t) = \int_D ((J*)^n * u)(x,t)dx,$$

for any $n \geq 1$, which is equivalent to what was obtained before.

As consequence of the above theorem, we obtain the following nonlocal version of (6.1) and (6.2).

Corollary 6.5 *Let u be the solution of (6.3) for the datum $u_0 = X_D$, with a finite Lebesgue measurable set D in $\mathbb{R}^N$, then*

$$\lim_{t \to 0^+} \frac{1}{t} \int_{\mathbb{R}^N \setminus D} u(x,t)dx = -(\mathbb{H}_D^J)'(0) = P_J(D), \qquad (6.18)$$

or equivalently,

$$\mathbb{H}_D^J(t) = |D| - P_J(D)t + o(t) \qquad as\ t \downarrow 0. \qquad (6.19)$$

Remark 6.6 Let us make a comment on the relation of the semigroup $T_J(t)$ and the operator $\Delta_J X_D$. Observe that for $\phi(x,t) = T_J(t)X_D(x) - X_D(x)$, by (6.4) we have

$$\int_{\mathbb{R}^N} \phi(x,t)dx = 0 \quad \forall t \geq 0,$$

hence

$$\int_{\mathbb{R}^N} \phi^+(x,t)dx = \int_{\mathbb{R}^N} \phi^-(x,t)dx \quad \forall t \geq 0.$$

Therefore, since

$$\phi^+(x,t) = (T_J(t)X_D(x) - X_D(x))X_{\mathbb{R}^N \setminus D}(x),$$

we have

$$\|T_J(t)X_D - X_D\|_{L^1} = 2 \int_{\mathbb{R}^N} \phi^+(x,t)dx = 2 \int_{\mathbb{R}^N \setminus D} T_J(t)X_D(x)dx.$$

Then, by (6.18), we get

$$\lim_{t \to 0^+} \frac{1}{2t} \|T_J(t)X_D - X_D\|_{L^1} = P_J(D). \qquad (6.20)$$

Note that this formula is similar to the one that holds for the classical heat content; see [61]. Now, since

$$P_J(D) = \frac{1}{2} \int_{\mathbb{R}^N} \int_{\mathbb{R}^N} J(x-y) \Big| X_D(y) - X_D(x) \Big| dy dx$$

$$= \frac{1}{2} \int_{\mathbb{R}^N} \Big| \int_{\mathbb{R}^N} J(x-y)(X_D(y) - X_D(x)) dy \Big| dx$$

$$= \frac{1}{2} \|\Delta_J X_D\|_{L^1},$$

we can write (6.20) as:

$$\lim_{t \to 0^+} \frac{1}{t} \|T_J(t) X_D - X_D\|_{L^1} = \|\Delta_J X_D\|_{L^1}. \tag{6.21}$$

Moreover, since the operator $B = \Delta_J$ is the infinitesimal generator of the C_0-semigroup $(T_J(t))_{t \geq 0}$ in $L^2(\mathbb{R}^N)$, we have

$$\Delta_J X_D = \lim_{t \to 0^+} \frac{1}{t}(T_J(t) X_D - X_D) \quad \text{in } L^2(\mathbb{R}^N). \tag{6.22}$$

Hence, (6.21) and (6.22) imply that also

$$\Delta_J X_D = \lim_{t \to 0^+} \frac{1}{t}(T_J(t) X_D - X_D) \quad \text{in } L^1(\mathbb{R}^N).$$

6.2.3   Nonlocal Curvature in the Expansion

We have proved that

$$\mathbb{H}_D^J(0) + 2(\mathbb{H}_D^J)'(0) + (\mathbb{H}_D^J)''(0) = \int_D \int_{\mathbb{R}^N} \int_D J(x - y)J(y - z)dzdydx. \tag{6.23}$$

Then, having in mind (6.9), we get

$$(\mathbb{H}_D^J)''(0) = \int_D \int_{\mathbb{R}^N} \int_D J(x - y)J(y - z)dzdydx + \int_D \mathcal{H}_{\partial D}^J(x)dx. \tag{6.24}$$

Therefore, the Taylor's expansion gives

$$\mathbb{H}_D^J(t) = |D| - P_J(D)t + \frac{1}{2}\int_D \mathcal{H}_{\partial D}^J(x)dx\, t^2$$

$$+ \frac{1}{2}\int_D \int_{\mathbb{R}^N} \int_D J(x - y)J(y - z)dzdydx\, t^2 + O(t^3) \text{ as } t \downarrow 0.$$

Observe that, for $n \geq 2$, we can express the coefficients $(\mathbb{H}_D^J)^{(n)}(0)$ by using the nonlocal curvature as follows:

$$(\mathbb{H}_D^J)^{(n)}(0) = (-1)^n P_J(D) +$$

$$+ \frac{1}{2}\sum_{k=1}^{n-1}\binom{n-1}{k}(-1)^{n-1-k}\left(\int_D \mathcal{H}_{\partial D}^{(J*)^k}(x)dx - \int_D \mathcal{H}_{\partial D}^{(J*)^{(k+1)}}(x)dx\right). \tag{6.25}$$

Indeed,

$$(\mathbb{H}_D^J)^{(n)}(0) = -\sum_{k=1}^{n} \binom{n}{k} (-1)^{n-k} P_{(J*)^k}(D)$$

$$= \sum_{k=1}^{n-1} (-1)^{n-1-k} \left(\binom{n-1}{k-1} + \binom{n-1}{k} \right) P_{(J*)^k}(D) - P_{(J*)^n}(D)$$

$$= (-1)^n P_J(D) + \sum_{k=1}^{n-1} \binom{n-1}{k} (-1)^{n-1-k} \left(P_{(J*)^k}(D) - P_{(J*)^{(k+1)}}(D) \right),$$

and hence, using (6.9), we can write

$$P_{(J*)^k}(D) - P_{(J*)^{(k+1)}}(D) = \frac{1}{2} \left(\int_D \mathcal{H}_{\partial D}^{(J*)^k}(x) dx - \int_D \mathcal{H}_{\partial D}^{(J*)^{(k+1)}}(x) dx \right)$$

to get (6.25).

6.2.4   A Probabilistic Interpretation

Let us show that the formula given in (6.15), that is,

$$\mathbb{H}_D^J(t) = \sum_{k=0}^{+\infty} \left(\int_D \int_D (J*)^k (x-y) dy dx \right) \frac{e^{-t} t^k}{k!},$$

has a probabilistic interpretation.

As mentioned after giving Definition 6.1 of the J-heat content, if $J(x - y)$ is thought of as the probability distribution of jumping from location y to location x, then

$$\int_{\mathbb{R}^N} J(x - s) J(s - y) ds$$

is the probability of jumping from location y to location x in two jumps (the probability of passing through a point s is $J(x - s) J(s - y)$ and we integrate for $s \in \mathbb{R}^N$). Further,

$$\int_{\mathbb{R}^N} \int_{\mathbb{R}^N} J(x - s) J(s - w) J(w - y) ds dw$$

is the probability of jumping from location y to location x in three jumps. Let us call these jumps as J-jumps.

Then, from $J(x - y)$ we obtain the probability of a transition from x to D in k steps or J-jumps as:

$$F^{(k)}(x, D) = \int_D (J*)^k(x - y)dy,$$

and we can also set

$$F^{(0)}(x, D) = \chi_D(x).$$

Observe that $F^{(k)}(x, D)$ also determines how many matter of D goes to x after k jumps, even for $k = 0$. Also, if we define

$$a(0, D) = \int_D \int_D (J*)^0(x - y)dydx = |D|,$$

and

$$a(k, D) = \int_D \int_D (J*)^k(x - y)dydx, \qquad k = 1, 2, \ldots,$$

we have that $a(k, D)$ is the amount of matter of D remaining in D after k jumps for any $k \geq 0$.

From (6.11), we have that $u(x, t)$ is the expected value of the amount of matter of D that goes to x when this matter moves by J-jumps, and the number of J-jumps up to time t, N_t, follows a Poisson distribution with rate t:

$$u(x, t) = \sum_{k=0}^{+\infty} F^{(k)}(x, D)\frac{e^{-t}t^k}{k!}.$$

It is well known that this function is the transition probability of a pseudo-Poisson process of intensity 1 (see [44, Ch. X]).

From (6.15), we have that

$$\mathbb{H}_D^J(t) = \sum_{k=0}^{+\infty} a(k, D)\frac{e^{-t}t^k}{k!}$$

is the expected value of the amount of matter of D that remains in D when this matter moves by J-jumps and the number of J-jumps up to time t follows a Poisson distribution with rate t. We can write it as:

$$\mathbb{H}_D^J(t) = \mathbb{E}(a(N_t, D)).$$

Let us decompose $f(k)$ for $k \geq 2$ as:

$$a(k, D) = b(k, D) + c(k, D)$$

where

$$b(k, D) = \int_D \int_{D^{k-1}} \int_D J(x - x_2) J(x_2 - x_3) \cdots J(x_k - y) dy dx_k \cdots dx_3 dx_2 dx$$

and

$$c(k, D) = \int_D \int_{\mathbb{R}^{N-1} \setminus D^{N-1}} \int_D J(x - x_2) J(x_2 - x_3) \cdots J(x_n - y) dy dx_n \cdots dx_3 dx_2 dx,$$

being $D^{n-1} = D \times D \times \cdots \times D$ $(n-1)$-times. Let us also define

$$b(0, D) = |D|$$

and

$$b(1, D) = \int_D \int_D J(x - y) dy dx.$$

We have that $b(k, D)$ represents the amount of matter of D remaining in D after k jumps all inside D. On the other hand, $c(k, D)$ represents the amount of matter of D remaining in D after k jumps with at least one outside D, of course this has only sense for $k \geq 2$. Set accordingly $c(0, D) = c(1, D) = 0$. Then, we can decompose the J-heat content as:

$$\mathbb{H}_D^J(t) = \mathbb{Q}_D^J(t) + \mathbb{R}_D^J(t),$$

where

$$\mathbb{Q}_D^J(t) = \sum_{k=0}^{+\infty} b(k, D) \frac{e^{-t} t^k}{k!}$$

and

$$\mathbb{R}_D^J(t) = \sum_{k=0}^{+\infty} c(k, D) \frac{e^{-t} t^k}{k!}.$$

The function $\mathbb{Q}_D^J(t)$ is the spectral J-heat content of D and will be studied in Sect. 6.4. We have that

$$\mathbb{Q}_D^J(t) < \mathbb{H}_D^J(t) \quad \text{for all } t > 0.$$

6.2.5   The J-heat Loss of D

Using (6.7), from (6.15) we can get the following expansion for the nonlocal J-heat loss of D in $\mathbb{R}^N$ at t (see [70]):

$$|D| - \mathbb{H}_D^J(t) = \sum_{k=0}^{+\infty} P_{(J*)^k}(D) \frac{e^{-t}t^k}{k!},$$

where we set $P_{(J*)^0}(D) = 0$. Using the above notation, we have

$$|D| - \mathbb{H}_D^J(t) = \mathbb{E}(g(N_t)),$$

where $g(k) = P_{(J*)^k}(D)$.

6.2.6   The J-heat Content and the Nonlocal Isoperimetric Inequality

From the isoperimetric inequality given in Theorem 2.1, we have the following result.

Corollary 6.7 *Let J be radially nonincreasing and having compact support $B_\delta(0)$. For any bounded subset $D \subset \mathbb{R}^N$ with $|D| > \frac{\delta}{2}$, we have*

$$\mathbb{H}_D^J(t) \leq \mathbb{H}_{B_r}^J(t) \quad \text{for small } t > 0,$$

where B_r is a ball such that $|B_r| = |D|$.

Proof The result is true when D is a ball of radius r, so let us suppose that this is not the case. Then, on account of point 2. of Remark 3.12, we have

$$\mathbb{H}_{B_r}^J(0) = |D| = \mathbb{H}_D^J(0)$$

and

$$(\mathbb{H}_{B_r}^J)'(0) = -P_J(B_r) > -P_J(D) = (\mathbb{H}_D^J)'(0).$$

Then, by (6.19), we get

$$\mathbb{H}_D^J(t) < \mathbb{H}_{B_r}^J(t) \quad \text{for small } t > 0,$$

and the result follows. $\square$

As consequence of Corollary 6.7 and Proposition 6.2, we have the following characterization.

Corollary 6.8 *Let J be radially nonincreasing and having compact support $B_\delta(0)$. The
Isoperimetric Inequality (2.2) is equivalent to the inequality:*

$$\|T_J(t)\chi_D\|_{L^2} \le \|T_J(t)\chi_{B_r}\|_{L^2} \quad \text{for small } t > 0,$$

with B_r being a ball such that $|B_r| = |D|$ when $r > \frac{\delta}{2}$.

A similar result for the local case was proved in [64] (see also [15] and [55]).

6.3   Convergence to the Heat Content when Rescaling the Kernel

For a subset D in $\mathbb{R}^N$ with finite Lebesgue measure, we will call

$$\mathbb{H}_D^{J,\alpha}(t) = \int_D u(x,t)dx$$

the J-heat content of α-intensity of D, where u is the solution of

$$\begin{cases} u_t(x,t) = \alpha \displaystyle\int_{\mathbb{R}^N} J(x-y)(u(y,t) - u(x,t))dy, & x \in \mathbb{R}^N, \ t \in [0,\infty), \\ \\ u(x,0) = \chi_D(x), & x \in \mathbb{R}^N. \end{cases}$$

Observe that $\mathbb{H}_D^J(t)$ is the J-heat content of 1-intensity of D; also,

$$\mathbb{H}_D^{J,\alpha}(t) = \mathbb{H}_D^J(\alpha t).$$

Let v_ϵ be the solution of

$$\begin{cases} (v_\epsilon)_t(x,t) = \dfrac{1}{\epsilon^2}[J_\epsilon * v_\epsilon(x,t) - v_\epsilon(x,t)], & x \in \mathbb{R}^N, \ t \in [0,\infty), \\ \\ v_\epsilon(x,0) = \chi_D(x), & x \in \mathbb{R}^N. \end{cases}$$

By Andreu-Vaillo et al. [12, Theorem 1.30], for J radially nonincreasing and compactly supported, we have

$$\lim_{\epsilon \to 0^+} \|v_\epsilon - v\|_{L^\infty(\mathbb{R}^N \times (0,T))} = 0$$

for every $T > 0$, with v being the solution of the heat equation:

$$\begin{cases} v_t(x, t) = \dfrac{1}{C_{J,2}} \Delta v(x, t), \; x \in \mathbb{R}^N, \; t \in [0, \infty), \\[2mm] v(x, 0) = \chi_D(x), \qquad\qquad x \in \mathbb{R}^N, \end{cases}$$

with

$$C_{J,2} = \frac{2}{\displaystyle\int_{\mathbb{R}^N} J(x)|x_N|^2 dx}. \tag{6.26}$$

Set now $u(x, t) = v(x, C_{J,2}t)$. Then, u verifies

$$\begin{cases} u_t(x, t) = \Delta u(x, t), \; x \in \mathbb{R}^N, \; t \in [0, \infty), \\[2mm] u(x, 0) = \chi_D(x), \qquad x \in \mathbb{R}^N. \end{cases}$$

Hence, for the solution u_ϵ of the problem:

$$\begin{cases} (u_\epsilon)_t(x, t) = \dfrac{C_{J,2}}{\epsilon^2} \left(J_\epsilon * u_\epsilon(x, t) - u_\epsilon(x, t) \right), \; x \in \mathbb{R}^N, \; t \in [0, \infty), \\[2mm] u_\epsilon(x, 0) = \chi_D(x), \qquad\qquad\qquad\qquad x \in \mathbb{R}^N, \end{cases}$$

we have that

$$\mathbb{H}_D^{J_\epsilon, \frac{C_{J,2}}{\epsilon^2}}(t) = \int_D u_\epsilon(x, t)\, dx = \int_D v_\epsilon\left(x, C_{J,2}t\right) dx,$$

and

$$\lim_{\epsilon \to 0} \int_D v_\epsilon\left(x, C_{J,2}t\right) dx = \int_D v\left(x, C_{J,2}t\right) dx = \int_D u(x, t)\, dx = \mathbb{H}_D(t).$$

Consequently, we have proved the following result:

Theorem 6.9 *Assume J is radially nonincreasing and compactly supported. For a subset D in $\mathbb{R}^N$ with finite Lebesgue measure, we have*

$$\lim_{\epsilon \to 0^+} \mathbb{H}_D^{J_\epsilon, \frac{C_{J,2}}{\epsilon^2}}(t) = \mathbb{H}_D(t) \quad \text{for all } t > 0.$$

That is, if the jumps are rescaled to occur in a ball of radius ϵ and the intensity of the Poisson process that controls the intensity of the jumps is rescaled to the size $\frac{C_J}{\epsilon^2}$, then we are approaching, for ϵ small, the Gaussian heat content.

Remark 6.10 In Theorem 1.11, it is shown that, under hypothesis (1.19) on J,

$$\lim_{\epsilon \downarrow 0} C_{J_\epsilon} P_{J_\epsilon}(E) = P(E), \tag{6.27}$$

for a bounded set $E \subset \mathbb{R}^N$ of finite perimeter.

Observe that

$$(J_\epsilon *)^k = [(J*)^k]_\epsilon.$$

Then, by Theorem 6.4, for $n \geq 1$, we have

$$(\mathbb{H}_D^{J_\epsilon})^{(n)}(0) = -\sum_{k=1}^{n} \binom{n}{k}(-1)^{n-k} P_{[(J*)^k]_\epsilon}(D),$$

and consequently, by (6.27), there holds

$$\lim_{\epsilon \to 0} \frac{1}{\epsilon}(\mathbb{H}_D^{J_\epsilon})^{(n)}(0) = c_n \operatorname{Per}(D),$$

where

$$\sum_{k=1}^{n} \binom{n}{k} c_k = \frac{1}{2} \int_{\mathbb{R}^N} (J*)^n(x)|x_N|dx,$$

that is,

$$c_n = -\frac{1}{2} \sum_{k=1}^{n} \binom{n}{k}(-1)^{n-k} \int_{\mathbb{R}^N} (J*)^k(x)|x_N|dx.$$

6.4   The Spectral Heat Content

The *spectral heat content* is given by:

$$\mathbb{Q}_D(t) = \int_D u(x, t)dx$$

for the solution u of the Dirichlet problem:

$$\begin{cases} u_t(x, t) = \Delta u(x, t), \ (x, t) \in D \times [0, \infty), \\[2mm] u(x, t) = 0, \qquad\qquad (x, t) \in \partial D \times (0, \infty), \\[2mm] u(x, 0) = \chi_D(x) \qquad x \in D. \end{cases}$$

The following result was given in [71] for smooth bounded domains:

$$\mathbb{Q}_D(t) = |D| - \frac{2}{\sqrt{\pi}}\text{Per}(D)\sqrt{t} + \frac{1}{2}(N-1)\int_{\partial D}\mathcal{H}_{\partial D}(x)d\mathcal{H}^{N-1}(x)\,t + O(t^{3/2}), \quad t \downarrow 0,$$

where $\mathcal{H}_{\partial D}$ is the mean curvature of ∂D.

In [12], the following nonlocal Dirichlet problem was also studied:

$$\begin{cases} u_t(x, t) = \int_{\mathbb{R}^N} J(x-y)(u(y, t) - u(x, t))dy, \ (x, t) \in D \times [0, \infty), \\[2mm] u(x, t) = 0, \qquad\qquad\qquad\qquad\qquad (x, t) \in (\mathbb{R}^N \setminus D) \times [0, \infty), \\[2mm] u(x, 0) = u_0(x) \qquad\qquad\qquad\qquad\qquad x \in \mathbb{R}^N. \end{cases} \tag{6.28}$$

Therefore, we can also define the spectral J-heat content as:

$$\mathbb{Q}_D^J(t) = \int_D u(x, t)dx, \tag{6.29}$$

with u being the solution of (6.28) for the datum $u_0 = \chi_D$. Observe that also

$$\mathbb{Q}_D^J(0) = |D|. \tag{6.30}$$

We will describe in this section which is the asymptotic expansion of $\mathbb{Q}_D^J(t)$. The first term in the expansion is given in (6.30). For the second term, it is easy to get again, from the derivative of (6.29), that

$$\mathbb{Q}_D^J(t) + (\mathbb{Q}_D^J)'(t) = \int_D \int_D J(x - y)u(y, t)dydx, \qquad (6.31)$$

consequently,

$$\mathbb{Q}_D^J(0) + (\mathbb{Q}_D^J)'(0) = \int_D \int_D J(x - y)dydx,$$

and hence,

$$(\mathbb{Q}_D^J)'(0) = -P_J(D).$$

Note that the first two terms in the expansion of $\mathbb{Q}_D^J(t)$ and of $\mathbb{H}_D^J(t)$ coincide. Now, the expression for the next terms differs from that of the J-heat content. Taking derivatives in (6.31), we get

$$\mathbb{Q}_D^J(t) + 2(\mathbb{Q}_D^J)'(t) + (\mathbb{Q}_D^J)''(t) = \int_D \int_D \int_D J(x - y)J(y - z)u(z, t)dzdydx. \quad (6.32)$$

Then, for example, now, instead of (6.23) we have

$$\mathbb{Q}_D^J(0) + 2(\mathbb{Q}_D^J)'(0) + (\mathbb{Q}_D^J)''(0) = \int_D \int_D \int_D J(x - y)J(y - z)dzdydx,$$

Hence,

$$(\mathbb{Q}_D^J)''(0) = \int_D \int_D \int_D J(x - y)J(y - z)dzdydx + 2P_J(D) - |D|,$$

that is,

$$(\mathbb{Q}_D^J)''(0) = \int_D \int_D \int_D J(x - y)J(y - z)dzdydx + \int_D \mathcal{H}_{\partial D}^J(x)dx,$$

and this term is different from (6.24) in the term with three integrals.

Gathering this information, we have the following result.

Theorem 6.11 *For $\mathbb{Q}_D^J(t)$ the J-spectral heat content of D, it holds that*

$$\mathbb{Q}_D^J(t) = |D| - P_J(D)\,t + \frac{1}{2}\int_D \mathcal{H}_{\partial D}^J(x)dx\,t^2$$

$$+ \frac{1}{2}\int_D\int_D\int_D J(x-y)J(y-z)dzdydx\,t^2 + O(t^3)\,as\,t \downarrow 0,$$

where $\mathcal{H}_{\partial D}^J(x)$ is the J-mean curvature at x.

Now, similarly to the J-heat content we also have the following complete expansion for the spectral J-heat content.

Theorem 6.12 *Let D be a finite measurable set. Then,*

$$\mathbb{Q}_D^J(t) = \sum_{n=0}^{+\infty} b(D,n)\frac{e^{-t}t^n}{n!} \qquad \forall t > 0,$$

where $b(0,D) = |D|$ and, for $n \geq 1$,

$$b(n,D) = \int_{D^{n+1}} J(x_1-x_2)J(x_2-x_3)\cdots J(x_n-x_{n+1})dx_{n+1}dx_n\cdots dx_3dx_2dx_1,$$

being $D^{n+1} = D \times D \times \cdots \times D,\ (n+1)$-times.

Proof Taking into account the procedure used to get (6.31) and (6.32), by induction, we obtain

$$\sum_{k=0}^{n} \binom{n}{k}(\mathbb{Q}_D^J)^{(k)}(t) = b(n,D)$$

for any $n \geq 1$. And from here,

$$\sum_{k=0}^{n} \binom{n}{k}(\mathbb{Q}_D^J)^{(k)}(0) = b(n,D) \qquad (6.33)$$

for any $n \geq 0$. Therefore using the Cauchy product of series and (6.33), we get that

$$e^t\,\mathbb{Q}_D^J(t) = \sum_{n=0}^{+\infty}\left(\sum_{k=0}^{n}\frac{1}{(n-k)!}\frac{(\mathbb{Q}_D^J)^{(k)}(0)}{k!}\right)t^n$$

$$= \sum_{n=0}^{+\infty} b(D,n)\frac{t^n}{n!}$$

for all $t > 0$. And, the proof is finished.   □

Set

$$\mathbb{Q}_D^{J,\alpha}(t) = \mathbb{Q}_D^{J}(\alpha t).$$

Following the same steps as in Sect. 6.3 and using now [12, Theorem 2.13], we obtain the following rescaling result between the spectral J-heat content and the spectral heat content.

Theorem 6.13 *Assume J is radially nonincreasing and compactly supported. For a bounded smooth domain D in $\mathbb{R}^N$, we have*

$$\lim_{\epsilon \to 0^+} \mathbb{Q}_D^{J_\epsilon, \frac{C_{J,2}}{\epsilon^2}} (t) = \mathbb{Q}_D(t) \quad \text{for all } t > 0,$$

where $C_{J,2}$ is given in (6.26).

6.5   Lack of Regularity for the Nonlocal Heat Diffusion

It is well known that there is no regularizing effect on the solutions $T(t)u_0$ of

$$\begin{cases} u_t(x,t) = \displaystyle\int_{\mathbb{R}^N} J(x-y)(u(y,t) - u(x,t))dy, \ (x,t) \in \mathbb{R}^N \times [0,+\infty), \\ u(x,0) = u_0(x), \qquad\qquad\qquad\qquad\qquad\qquad x \in \mathbb{R}^N. \end{cases}$$

We will see in this section some facts on this question.

Let D be a finite measurable set. Let $u(\cdot, t) = T_J(t)\chi_D$. Using (6.8), we can express (6.11) as follows:

$$u(x,t) = \frac{1}{2} - \frac{1}{2}e^{-t} \sum_{n=0}^{\infty} H_{\partial D}^{(J*)^n}(x)\frac{t^n}{n!}, \tag{6.34}$$

where $H_{\partial D}^{(J*)^0}(x) = 1 - 2\chi_D(x)$.

Proposition 6.14 *Let D be a finite measurable set. Let $u(\cdot, t) = T_J(t)\chi_D$. Given any level α between 0 and 1, then*

$$\{x \in \mathbb{R}^N : u(x,t) > \alpha\} = D \quad \text{for all } 0 \le t < \min\{\ln(1/\alpha), \ln(1/(1-\alpha))\}.$$

Proof From (6.34), we have

$$\left\{x \in \mathbb{R}^N : u(x,t) > \alpha\right\} = \left\{x \in \mathbb{R}^N : \sum_{n=0}^{\infty} H_{\partial D}^{(J*)^n}(x)\frac{t^n}{n!} < (1 - 2\alpha)e^t\right\}$$

$$= \left\{x \in D : \sum_{n=1}^{\infty} H_{\partial D}^{(J*)^n}(x)\frac{t^n}{n!} < 1 + (1 - 2\alpha)e^t\right\}$$

$$\cup \left\{x \in \mathbb{R}^N \setminus D : \sum_{n=1}^{\infty} H_{\partial D}^{(J*)^n}(x)\frac{t^n}{n!} < -1 + (1 - 2\alpha)e^t\right\}.$$

Now, since $-1 \le H_{\partial D}^{(J*)^n} \le 1$, we have that

$$-(e^t - 1) \le \sum_{n=1}^{\infty} H_{\partial D}^{(J*)^n}(x)\frac{t^n}{n!} \le e^t - 1.$$

Then, on one hand,

$$e^t - 1 < 1 + (1 - 2\alpha)e^t \quad \Leftrightarrow \quad t < \ln(1/\alpha)$$

Hence, for $t < \ln(1/\alpha)$,

$$\sum_{n=1}^{\infty} H_{\partial D}^{(J*)^n}(x)\frac{t^n}{n!} < 1 + (1 - 2\alpha)e^t.$$

On the other hand,

$$-(e^t - 1) > -1 + (1 - 2\alpha)e^t \quad \Leftrightarrow \quad t < \ln(1/(1 - \alpha)).$$

Hence, for $t \le \ln(1/(1 - \alpha))$,

$$\sum_{n=1}^{\infty} H_{\partial D}^{(J*)^n}(x)\frac{t^n}{n!} > -1 + (1 - 2\alpha)e^t.$$

Consequently, for $t < \min\{\ln(1/\alpha), \ln(1/(1 - \alpha))\}$, we have that

$$\left\{x \in \mathbb{R}^N : u(x,t) > \alpha\right\} = D.$$

$\square$

Theorem 6.15 *Let D be a bounded open set and $u(\cdot, t) = T_J(t)\chi_D$. For any $x \in \partial D$,*

$$u(x^-, t) - u(x^+, t) = e^{-t},$$

where

$$u(x^-, t) = \lim_{y \to x,\, y \in D} u(y, t)$$

and

$$u(x^+, t) = \lim_{y \to x,\, y \notin D} u(y, t).$$

That is, for any time t there is a jump on the boundary ∂D in the solution of (6.3). This jump has the same height e^{-t} at every point.

Proof From (6.34), since $H_{\partial D}^{(J*)^0}(x) = 1 - 2\chi_D(x)$, and the continuity of the curvature, we have

$$u(x^-, t) = \frac{1}{2} + \frac{1}{2}e^{-t} - \frac{1}{2}e^{-t} \sum_{n=1}^{\infty} H_{\partial D}^{(J*)^n}(x) \frac{t^n}{n!}$$

and

$$u(x^+, t) = \frac{1}{2} - \frac{1}{2}e^{-t} - \frac{1}{2}e^{-t} \sum_{n=1}^{\infty} H_{\partial D}^{(J*)^n}(x) \frac{t^n}{n!}.$$

Hence, the result is straightforward.   □

The initial velocities on the solution depend on the nonlocal J-curvature at each point. This follows directly from (6.5).

Proposition 6.16 *Let D be a finite measurable set and $u(\cdot, t) = T_J(t)\chi_D$. Then,*

$$u_t(x, 0) = -\frac{1}{2}(1 + H_{\partial D}^J(x)) \leq 0 \qquad \text{for } x \in D,$$

$$u_t(x, 0) = \frac{1}{2}(1 - H_{\partial D}^J(x)) \geq 0 \qquad \text{for } x \notin D.$$

This implies that

$$\{x : H_{\partial D}^J(x) < 0\} = \left\{x : |u_t(x, 0)| > \frac{1}{2}\right\}.$$

A Nonlocal Mean Curvature Flow

Consider a family $\{\Gamma_t\}_{t\geq 0}$ of hypersurfaces embedded in $\mathbb{R}^N$ parametrized by time t. Assume that each $\Gamma_t = \partial E_t$, the boundary of a bounded open set E_t in $\mathbb{R}^N$. A *geometric flow* is an evolution for the sets $t \mapsto E_t$ governed by a law of the form:

$$V(x, t) = -K(x, E_t), \tag{7.1}$$

called surface evolution equation, where $V(x, t)$ stands for the (outer) normal velocity of ∂E_t at x and $K(\cdot, E)$ is some generalized curvature of ∂E.

If the function $K(x, E)$ depends only on how ∂E looks around x, the flow is called local, this is the case of the classical *mean curvature flow*, where $K(x, E) = H_{\partial E}(x)$ is the mean curvature of ∂E at x. Now, for some relevant geometric flows, $K(x, E)$ is truly nonlocal and depends on the global shape of the evolving set E_t itself. It happens for instance for *fractional mean curvature flows*, where

$$K(x, E) = \lim_{\delta \downarrow 0} s(1 - s) \int_{\mathbb{R}^N \setminus B_\delta(x)} \frac{\chi_{\mathbb{R}^N \setminus E}(y) - \chi_E(y)-}{|x - y|^{N+s}} dy,$$

that has been recently studied by Saez and Valdinoci [65].

In [32], Chambolle, Morini and Ponsiglione study nonlocal curvature flows of the form (7.1) for some general class of generalized curvatures.

Let $J : \mathbb{R}^N \to [0, +\infty[$ be a measurable, nonnegative and radially nonincreasing symmetric function verifying

$$\int_{\mathbb{R}^N} J(z)dz = 1.$$

© Springer Nature Switzerland AG 2019

J. M. Mazón et al., *Nonlocal Perimeter, Curvature and Minimal Surfaces
for Measurable Sets*, Frontiers in Mathematics,
https://doi.org/10.1007/978-3-030-06243-9_7

In Chap. 3, associated with J, it has been studied the J-curvature of a measurable set $E \subset \mathbb{R}^N$:

$$H_{\partial E}^J(x) = -\int_{\mathbb{R}^N} J(x - y)(\chi_E(y) - \chi_{\mathbb{R}^N \setminus E}(y)) dy = 1 - 2 \int_E J(x - y) dy.$$

In the case that J is continuous, $K(x, E) = H_{\partial E}^J(x)$ is one of the generalized curvatures considered in [32] and therefore the nonlocal mean curvature flow:

$$V(x, t) = -H_{\partial E_t}^J(x). \tag{7.2}$$

As consequence of the results in [32], it is well known the level set formulation of (7.2) and also it is studied a generalized Almgren–Taylor–Wang minimizing movements scheme to approximate the geometric motion associated with (7.2).

It is not our aim here to continue with the study of problem (7.2) but to state a similar problem in which also the normal velocity is nonlocal and has a probabilistic interpretation. Note that in the general problem (7.1), the normal velocity $V(x, t)$ is given by the velocity in the direction of the (outer) unitary normal vector of Γ_t at x, that we will denote as $v_{\partial E}(x)$, which implies to assume some smoothness of Γ_t. To define the nonlocal normal vector, we do not need any regularity of the set.

7.1   Nonlocal Normal Vector

We define the nonlocal normal vector in the following way.

Definition 7.1 Given a measurable set $E \subset \mathbb{R}^N$, we define the *J-normal (inward) vector* as:

$$v_E^J(x) = \int_E J(x - y)(y - x)\, dy = \left(\int_E J(x - y)(y_i - x_i)\, dy \right)_{i=1}^{i=N} \qquad \text{for } x \in \mathbb{R}^N.$$

Observe that this vector, as the J-curvature, is defined for any point, not necessarily for points on the boundary.

If we assume that we are working with a homogeneous population with density 1, and that $J(x - y)$ is the probability that an individual at y jumps to x, then

$$\int_E J(x - y) dy$$

gives the rate of individuals leaving x for going to any other site in E (or the rate of individuals going from E onto x), and

$$\frac{1}{\int_E J(x-y)dy} \int_E J(x-y)ydy$$

gives the expected place where individuals move from x onto E. As an expected value, it can be not a value in E. Then, the J-normal vector that can be written as:

$$v_E^J(x) = \int_E J(x-y)dy \left(\frac{1}{\int_E J(x-y)dy} \int_E J(x-y)y\,dy - x \right),$$

is *the expected direction* that individuals take when they move from x onto E following the law J, modulated in such a way that

$$v_E^J(x) = -v_{\mathbb{R}^N \setminus E}^J(x).$$

Definition 7.2 We say that a measurable set $E \subset \mathbb{R}^N$ is *J-regular* if $v_E^J(x) \neq 0$ for all $x \in \partial E$.

Definition 7.3 For a J-regular set E, we define the *unit (inward) J-normal vector* as:

$$v_{\partial E}^J(x) = \frac{v_E^J(x)}{|v_E^J(x)|}, \quad x \in \partial E.$$

The notation is similar but the unit normal has ∂E as subindex. Observe that

$$v_{\partial E}^J(x) = -v_{\partial (\mathbb{R}^N \setminus E)}^J(x).$$

Example 7.4 We claim that

$$v_{B_r(0)}^J(x) = -C(J,r)\frac{x}{r} \quad \text{for } x \in \partial B_r(0), \tag{7.3}$$

where $C(J,r)$ is the following positive constant (observe that $w_1 > 0$ in $B_r(r\mathbf{e}_1)$):

$$C(J,r) := \int_{B_r(r\mathbf{e}_1)} J(w)w_1 dw.$$

In fact,

$$v_{B_r(0)}^J(x) = \int_{B_r(0)} J(x-y)(y-x)\,dy = -\int_{B_r(x)} J(z)z\,dz.$$

Let R_x be the rotation in $\mathbb{R}^N$ such that $R_x(r\mathbf{e}_1) = x$, and assume that $(a_{i,j})$ is the matrix associated to the rotation R_x. Then, changing variables, using that J is radial, we obtain

$$\int_{B_r(x)} J(z)z_i \, dz = \int_{B_r(r\mathbf{e}_1)} J(R_x(w))R_x(w)_i dw = \int_{B_r(r\mathbf{e}_1)} J(w)R_x(w)_i dw$$

$$= \sum_{j=1}^{N} a_{i,j} \int_{B_r(r\mathbf{e}_1)} J(w)w_j dw = a_{i,1} \int_{B_r(r\mathbf{e}_1)} J(w)w_1 dw.$$

Now, since $R_x(r\mathbf{e}_1) = x$, we have $x = r(a_{1,1}, a_{2,1}, \ldots, a_{N,1})$. Therefore,

$$\int_{B_r(x)} J(z)z_i \, dz = \frac{x_i}{r} \int_{B_r(r\mathbf{e}_1)} J(w)w_1 dw$$

and we conclude the proof of (7.3).

Example 7.5 Let us now compute this nonlocal normal vector for a set with nonsmooth boundary. Consider the square $E = \text{conv}\{(0,0), (2,0), (2,2), (0,2)\}$, and $J = \frac{1}{|B_1(0)|}\chi_{B_1(0)}$. Then, for $0 \le s \le 1$, we have

$$v_E^J(s,0) := \left(\int_E J((s,0) - (y_1, y_2))(y_i - x_i) \, dy \right)_{i=1}^{i=2}$$

$$= \frac{1}{\pi} \left(\int_E \chi_{B_1((s,0))}(y_1, y_2)(y_1 - s) \, dy, \int_E \chi_{B_1((s,0))}(y_1, y_2)y_2 \, dy \right)$$

$$= \frac{1}{6\pi} \left(2(1 - s^2)^{3/2}, \, 2 + 3s - s^3 \right).$$

Note that the classical normal vector is not defined on $(0,0)$, but

$$v_{\partial E}^J(0,0) = \left(\frac{1}{\sqrt{2}}, \frac{1}{\sqrt{2}} \right).$$

Also,

$$v_{\partial E}^J(s,0) \ne (0,1) = -v_{\partial E}(s,0), \quad \text{if } 0 < s < 1.$$

Observe that for this set,

$$H_{\partial E}^J(s,0) = 1 - \frac{2}{\pi} \int_E \chi_{B_1((s,0))}(y)dy = \frac{\pi}{2} - \arcsin(s) - s\sqrt{1 - s^2}.$$

7.2   Nonlocal Mean Curvature Flow

We consider a J-regular set E_0 and we are interested in a family of J-regular sets E_t that satisfies, for every point $x \in \partial E_0$, the law of motion:

$$\begin{cases} \dfrac{\partial X}{\partial t}(x,t) = C_J H^J_{\partial E_t}(X(x,t)) v^J_{\partial E_t}(X(x,t)), & t > 0, \\[2mm] X(x,0) = x \in \partial E_0, \end{cases} \tag{7.4}$$

where $X(x,t)$ parameterizes the point of the boundary ∂E_t where x has moved after a time t. The constant $C_J = 2 \left(\int_{\mathbb{R}^N} J(z)|z_N| dz \right)^{-1}$ has been introduced in previous chapters, it is a rescaling constant whose role appears later. Problem (7.4) can be stated for any other positive constant.

In terms of the probabilistic description given above, in (7.4) we are describing the movement of $X(x,t)$ following *the nonlocal mean curvature flow* given by *the expected direction* $v^J_{\partial E_t}(X(x,t))$ times the push of the curvature $H^J_{\partial E_t}(X(x,t))$, which is proportional to the difference between the rate of individuals arriving to $X(x,t)$ from E_t and the rate of individuals moving to $X(x,t)$ from $\mathbb{R}^N \setminus E_t$ under the law J. Therefore, when the curvature is positive on the point $X(x,t)$, the movement will point toward E_t, in an expectation sense, and the contrary if the curvature is negative; in this last case, we can look at the flow as:

$$H^J_{\partial E_t}(X(x,t)) v^J_{\partial E_t}(X(x,t)) = -H^J_{\partial E_t}(X(x,t)) v^J_{\partial(\mathbb{R}^N \setminus E_t)}(X(x,t)).$$

Note that if E_t is a solution of (7.4), then E_t is also a solution of

$$\begin{cases} \frac{\partial X}{\partial t}(x,t) \cdot v^J_{\partial E_t}(X(x,t)) = C_J H^J_{\partial E_t}(X(x,t)), & t > 0, \\[2mm] X(x,0) = x \in \partial E_0, \end{cases} \tag{7.5}$$

If we write

$$V_J(x,t) = \frac{\partial X}{\partial t}(x,t) \cdot v^J_{\partial E_t}(X(x,t))$$

for the *J-normal velocity* of E_t at $X(x,t)$ and at time t in the direction of $v^J_{\partial E_t}(X(x,t))$, then the geometrical evolution problem (7.5) can be written as:

$$V_J = C_J H^J_{\partial} v^J_{\partial}.$$

Let us see that a circumference that evolves by nonlocal mean curvature collapses in finite time as it does when it evolves by (local) mean curvature.

Example 7.6 Suppose that $N = 2$ and $J = \frac{1}{|B_r(0)|}\chi_{B_r(0)}$. Let $E_0 = B_R(0)$ and assume that the boundary of E_0 evolves to the boundary of $E_t = B_{R(t)}(0)$ by nonlocal mean curvature following (7.4).

By Example 3.6, we have

$$H^J_{\partial E_t}(X(x,t)) = 1 - \frac{2|B_r(X(x,t)) \cap B_{R(t)}(0)|}{|B_r(0)|},$$

which is always positive. On the other hand, by Example 7.4, we have

$$v^J_{E_t}(X(x,t)) = -\int_{B_{R(t)}(R(t)e_1)} J(w)w_1 dw \frac{X(x,t)}{R(t)}$$

$$= -\frac{1}{\pi r^2}\int_{B_{R(t)}(R(t)e_1) \cap B_r(0)} w_1 dw \frac{X(x,t)}{R(t)}.$$

Therefore, in this case Problem (7.4) is

$$\begin{cases} \frac{\partial X}{\partial t}(x,t) = -F(R(t))\frac{X(x,t)}{R(t)}, & t > 0, \\[2mm] X(x,0) = x \in \partial E_0, \end{cases}$$

with

$$F(R(t)) := C_J \left(1 - \frac{2|B_r((R(t),0)) \cap B_{R(t)}(0)|}{|B_r(0)|}\right)$$

$$= \frac{3\pi}{2r}\left(1 - \frac{2|B_r((R(t),0)) \cap B_{R(t)}(0)|}{|B_r(0)|}\right).$$

Hence,

$$X(x,t) = x \exp\left(-\int_0^t F(R(s))ds\right),$$

and

$$R(t) = |X(x,t)| = R \exp\left(-\int_0^t F(R(s))ds\right).$$

Consequently, $R(t)$ satisfies

$$\begin{cases} R'(t) = -F(R(t)), & t > 0, \\ \\ R(0) = R, \end{cases}$$

and it is a nonincreasing function of t.

Assume $r \le 2R$, and let

$$T_r = \inf\{t \ge 0 : 2R(t) = r\}.$$

Then, $r \le 2R(t)$ for all $0 \le t \le T_r$. Let $Q(s) = R(s)^{-1}$. Then, for $0 \le t \le T_r$, we have

$$t = Q(R(t)) - Q(R) = -\int_{R(t)}^{R} Q'(R(s))ds = \int_{R(t)}^{R} \frac{1}{F(R(s))}ds,$$

hence,

$$T_r = \int_{\frac{r}{2}}^{R} \frac{1}{F(R(s))}ds.$$

Now, for $0 \le t \le T_r$, we have

$$|B_r((R(t), 0)) \cap B_{R(t)}(0)|$$

$$= r^2 \cos^{-1}\left(\frac{r}{2R(t)}\right) + R(t)^2 \cos^{-1}\left(1 - \frac{r^2}{2R(t)^2}\right) - \frac{1}{2}\sqrt{r^3(2R(t) + r)},$$

and then,

$$F(R(t)) =$$

$$1 - \frac{1}{\pi r^2}\left(r^2 \cos^{-1}\left(\frac{r}{2R(t)}\right) + R(t)^2 \cos^{-1}\left(1 - \frac{r^2}{2R(t)^2}\right) - \frac{1}{2}\sqrt{r^3(2R(t) + r)}\right).$$

On the other hand, since $R(t) \le \frac{r}{2}$ for all $t > T_r$, we have

$$F(R(t)) = \frac{3\pi}{2r}\left(1 - \frac{2R(t)^2}{r^2}\right) \qquad \text{for all } t > T_r.$$

Consequently, $R(t)$ satisfies

$$
\begin{cases}
R'(t) = \frac{3R(t)^2}{r^3} - \frac{3\pi}{2r}, & \text{for all } t > T_r, \\[2ex]
R(T_r) = \frac{r}{2}.
\end{cases}
$$

whose general solution is given by:

$$
\log\left(\frac{r\sqrt{\pi} - R(t)}{r\sqrt{\pi} + R(t)}\right) = \frac{6\sqrt{\pi}}{r^2}t + C \quad \text{for all } t \geq T_r.
$$

Since $R(T_r) = \frac{r}{2}$, we get

$$
C = \log\left(\frac{2\sqrt{\pi} - 1}{2\sqrt{\pi} + 1}\right) - \frac{6\sqrt{\pi}}{r^2}T_r,
$$

and therefore,

$$
\log\left[\left(\frac{r\sqrt{\pi} - R(t)}{r\sqrt{\pi} + R(t)}\right)\left(\frac{2\sqrt{\pi} + 1}{2\sqrt{\pi} - 1}\right)\right] = \frac{6\sqrt{\pi}}{r^2}(t - T_r) \quad \text{for all } t \geq T_r.
$$

From here,

$$
\left(\frac{2\sqrt{\pi} - 1}{2\sqrt{\pi} + 1}\right)\exp\left(\frac{6\sqrt{\pi}}{r^2}(t - T_r)\right) = \frac{r\sqrt{\pi} - R(t)}{r\sqrt{\pi} + R(t)}
$$

and this collapses in finite time

$$
T_r^e = T_r + \frac{r^2}{6\sqrt{\pi}}\log\left(\frac{2\sqrt{\pi} + 1}{2\sqrt{\pi} - 1}\right). \tag{7.6}
$$

For the (local) mean curvature flow, the circle collapses at time $R^2/2$ since in this case $R(t)$ is given by $\sqrt{R^2 - 2t}$. Observe that if we take $r \downarrow 0$ in (7.6) we obtain also that collapsing time:

$$
\lim_{r \downarrow 0} T_r^e = \lim_{r \downarrow 0} \int_{\frac{r}{2}}^{R} \frac{1}{F(R(s))}ds = \int_0^R s\, ds = \frac{R^2}{2}.
$$

Consider the rescaled kernel $J_\epsilon(x) = \frac{1}{\epsilon^N}J\left(\frac{x}{\epsilon}\right)$ for $\epsilon > 0$, yet introduced in previous chapters. In Example 7.4, we have seen that

$$
v_{B_r(0)}^J(x) = -C(J, r)\frac{x}{r} \quad \text{for } x \in \partial B_r(0),
$$

where $C(J, r) = \int_{B_r(re_1)} J(w)w_1 dw$. Now,

$$C_{J_\epsilon} C(J_\epsilon, r) = \frac{1}{\epsilon} C_J \frac{1}{\epsilon^N} \int_{B_r(re_1)} J\left(\frac{w}{\epsilon}\right) w_1 dw = \frac{2}{\int_{\mathbb{R}^N} J(z)|z_N| dz} \int_{B_{\frac{r}{\epsilon}}(\frac{r}{\epsilon} e_1)} (z) z_1 dz,$$

and consequently, $C_{J_\epsilon} C(J_\epsilon, r) \to 1$ as $\epsilon \to 0$. Therefore,

$$\lim_{\epsilon \downarrow 0} C_{J_\epsilon} v_{B_r(0)}^{J_\epsilon}(x) = -\frac{x}{r} = -v_{B_r(0)}(x).$$

Let us see that this convergence result is true for smooth domains.

Theorem 7.7 *Let $E \subset \mathbb{R}^N$ be a smooth set such that ∂E is of class C^2. Then, for every $x \in \partial E$, we have*

$$\lim_{\epsilon \downarrow 0} C_{J_\epsilon} v_E^{J_\epsilon}(x) = -v_{\partial E}(x), \tag{7.7}$$

where $v_{\partial E}(x)$ is the outward unit normal vector at $x \in \partial E$. And, for the unit J-normal vector, we have

$$\lim_{\epsilon \downarrow 0} v_{\partial E}^{J_\epsilon}(x) = -v_{\partial E}(x),$$

Proof We can assume that $x = 0 \in \partial E$. Namely, suppose ∂E is described as a graph in normal coordinates, meaning that, in an open ball B_{r_0}, ∂E coincides with the graph of a C^2 function $\varphi : B_{r_0} \cap \mathbb{R}^{N-1} \to \mathbb{R}$ with $\varphi(0) = 0$ and $\nabla \varphi(0) = 0$ such that $E \cap B_{r_0} = \{(y_1, \ldots, y_N) : y_N < \varphi(y_1, \ldots, y_{N-1})\}$. Then, to prove (7.7), we show that

$$\lim_{\epsilon \downarrow 0} C_{J_\epsilon} \int_E J_\epsilon(y) y \, dy = (0, 0, \ldots, 0, -1) = -v_{\partial E}(0).$$

For $\epsilon > 0$ such that $\epsilon r \leq r_0$, we have

$$C_{J_\epsilon} \int_E J_\epsilon(y) y \, dy = \frac{C_{J_\epsilon}}{\epsilon^N} \int_{E \cap B_{\epsilon r}(0)} J\left(\frac{y}{\epsilon}\right) y \, dy$$

$$= \frac{C_J}{\epsilon^{N+1}} \int_{\{y_N < \varphi(y_1, \ldots, y_{N-1})\} \cap B_{\epsilon r}(0)} J\left(\frac{y}{\epsilon}\right) y \, dy.$$

Changing variables as $z = \frac{y}{\epsilon}$, we have

$$\frac{C_J}{\epsilon} \int_E J_\epsilon(y) y \, dy = C_J \int_{\{z_N < \frac{1}{\epsilon}\varphi(\epsilon z_1, \ldots, \epsilon z_{N-1})\} \cap B_r(0)} J(z) z \, dz$$

On the other hand, by Taylor's expansion, we have

$$\varphi(\epsilon z_1, \ldots, \epsilon z_{N-1}) = \frac{1}{2} D^2 \varphi(0)(\epsilon z_1, \ldots, \epsilon z_{N-1}) + O(\epsilon^3) = \frac{1}{2} \sum_{i=1}^{N-1} \lambda_i \epsilon^2 z_i^2 + O(\epsilon^3),$$

being $\lambda_1, \ldots, \lambda_{N-1}$ the real eigenvalues of the symmetric $D^2 \varphi(0)$. Then, we have

$$\lim_{\epsilon \to 0} \frac{C_J}{\epsilon} \int_E J_\epsilon(y) y \, dy = \lim_{\epsilon \to 0} C_J \int_{\{z_N < \epsilon \frac{1}{2} \sum_{i=1}^{N-1} \lambda_i z_i^2\} \cap B_r(0)} J(z) z \, dz$$

$$= C_J \int_{\{z_N < 0\}} J(z) z \, dz = -1,$$

since, by symmetry,

$$\int_{\{z_N < 0\}} J(z) z_i \, dz = 0 \quad \forall i \neq N,$$

and

$$\int_{\{z_N < 0\}} J(z) z_N \, dz = -\frac{1}{2} \int_{\mathbb{R}^N} J(z) |z_N| \, dz = -\frac{1}{C_J}.$$

$\square$

In Theorem 3.7, assuming J is continuous at 0, we have proved that if $E \subset \mathbb{R}^N$ is a smooth set such that ∂E is of class C^2, then, for every $x \in \partial E$, we have

$$\lim_{\epsilon \downarrow 0} C_{J_\epsilon} H_{\partial E}^{J_\epsilon}(x) = (N - 1) H_{\partial E}(x),$$

where $H_{\partial E}(x)$ is the (local) mean curvature of ∂E at x. Therefore, as consequence of Theorems 3.7 and 7.7, we have the following result.

Corollary 7.8 *Assume J is continuous at 0. Let $E \subset \mathbb{R}^N$ be a smooth set such that ∂E is of class C^2. Then,*

$$\lim_{\epsilon \downarrow 0} C_{J_\epsilon} H_{\partial E}^{J_\epsilon}(x) v_{\partial E}^{J_\epsilon}(x) = -(N - 1) H_{\partial E}(x) v_{\partial E}(x) \qquad \forall x \in \partial E.$$

The above result motivates us to say that

$$\begin{cases} \dfrac{\partial X}{\partial t}(x, t) = C_{J_\epsilon} H_{\partial E_t^J}^J(X(x, t)) v_{\partial E_t^J}^J(X(x, t)), & t > 0, \\ X(x, 0) = x \in \partial E_0, \end{cases}$$

is a nonlocal counterpart of the classical mean curvature flow problem given by:

$$\begin{cases} \dfrac{\partial X}{\partial t}(x,t) = -(N-1)H_{\partial E_t}(X(x,t))\nu_{\partial E_t}(X(x,t)), & t > 0, \\[2ex] X(x,0) = x \in \partial E_0. \end{cases}$$

Let $\mathfrak{C}$ be the class of subsets of $\mathbb{R}^N$ that can be obtained as the closure of an open set with compact C^2 boundary. In [32], it is said that a curvature $K(x,E)$, defined for $E \in \mathfrak{C}$ and $x \in \partial E$, is the *first variation of the perimeter P_J* if for every $E \in \mathfrak{C}$, and any one-parameter family of diffeomorphisms (Φ_ϵ) of class C^2 both in x and ϵ with $\Phi_0(x) = x$, it holds

$$\frac{d}{d\epsilon}P_J(\Phi_\epsilon(E))|_{\epsilon=0} = \int_{\partial E} K(x,E)\psi(x) \cdot \nu_{\partial E}(x) d\mathcal{H}^{N-1}(x),$$

where $\psi(x) := \frac{\partial \Phi_\epsilon}{\partial \epsilon}|_{\epsilon=0}$.

Proposition 7.9 *If J is continuous, then the J-curvature is the first variation of the J-perimeter.*

Proof Since J is continuous, it is easy to see that $H^J_{\partial E}$ satisfies the continuity property (C) of [32]. Then by Chambolle et al. [32, Proposition 4.5], to proof the result we need to show that given $\varphi \in C^2(\mathbb{R}^N)$, and any $t_1 < t_2$ such that $D\varphi \neq 0$ in the set $\{t_1 \leq \varphi \leq t_2\}$, one has

$$P_J(\{\varphi \geq t_1\}) = P_J(\{\varphi \geq t_2\}) + \int_{\{t_1 < \varphi < t_2\}} H^J_{\{\varphi \geq \varphi(x)\}}(x)dx. \tag{7.8}$$

Now,

$$\int_{\{t_1 < \varphi < t_2\}} H^J_{\{\varphi \geq \varphi(x)\}}(x)dx$$

$$= \int_{\mathbb{R}^N} \chi_{\{t_1 < \varphi < t_2\}}(x) \int_{\mathbb{R}^N} (\chi_{\mathbb{R}^N \setminus \{t_1 < \varphi < t_2\}}(y) - \chi_{\{t_1 < \varphi < t_2\}}(y)) J(x-y) dy dx$$

$$= \frac{1}{2} \int_{\mathbb{R}^N \times \mathbb{R}^N} (\chi_{\{t_1 < \varphi < t_2\}}(x) - \chi_{\{t_1 < \varphi < t_2\}}(y)) \times$$

$$(\chi_{\mathbb{R}^N \setminus \{t_1 < \varphi < t_2\}}(y) - \chi_{\{t_1 < \varphi < t_2\}}(y)) J(x-y) dy dx$$

$$= \frac{1}{2} \int_{\mathbb{R}^N \times \mathbb{R}^N} |\chi_{\{t_1 < \varphi < t_2\}}(x) - \chi_{\{t_1 < \varphi < t_2\}}(y)| J(x-y) dy dx.$$

Then, since

$$|\chi_{\{t_1 < \varphi < t_2\}}(x) - \chi_{\{t_1 < \varphi < t_2\}}(y)|$$

$$= |\chi_{\{\varphi \ge t_1\}}(x) - \chi_{\{\varphi \ge t_1\}}(y)| - |\chi_{\{\varphi \ge t_2\}}(x) - \chi_{\{\varphi \ge t_3\}}(y)|,$$

we arrive to

$$\int_{\{t_1 < \varphi < t_2\}} H_{\{\varphi \ge \varphi(x)\}}^J(x) dx$$

$$= \frac{1}{2} \int_{\mathbb{R}^N \times \mathbb{R}^N} |\chi_{\{\varphi \ge t_1\}}(x) - \chi_{\{\varphi \ge t_1\}}(y)| J(x-y) dy dx$$

$$- \frac{1}{2} \int_{\mathbb{R}^N \times \mathbb{R}^N} |\chi_{\{\varphi \ge t_2\}}(x) - \chi_{\{\varphi \ge t_2\}}(y)| J(x-y) dy dx$$

$$= P_J(\{\varphi \ge t_1\}) - P_J(\{\varphi \ge t_2\}),$$

and (7.8) holds.   □

Remark 7.10 Assume that J is C^∞ and suppose E_t is a smooth solution of (7.4). Then applying [45, Theorem 6.1], we have

$$\frac{d}{dt} P_J(E_t) = \int_{\partial E_t} \left(H_{\partial E_t}^J(x) \right)^2 v_{\partial E_t}^J(x) \cdot v_{\partial E_t}(x) \, d\mathcal{H}^{N-1}(x).$$

Therefore, the J-perimeter of the evolving sets E_t decreases if $v_{\partial E_t}^J(x) \cdot v_{E_t}(x) \le 0$ a.e $-$ $\mathcal{H}^{N-1}$. In the case that $E_t = B_{R(t)}(0)$, by Example 7.4, we have that

$$v_{\partial B_{R(t)}(0)}^J(x) = -\frac{x}{R(t)} \quad \text{for } x \in \partial B_{R(t)}(0).$$

Therefore,

$$\frac{d}{dt} P_J(B_{R(t)}) = -\int_{\partial B_{R(t)}} \left(H_{\partial B_{R(t)}}^J(x) \right)^2 d\mathcal{H}^{N-1}(x).$$

Consequently, without solving the nonlocal mean curvature flow, we can state that the J-perimeters of the evolving balls decrease in time.

Bibliography

1. N. Abatangelo, E. Valdinoci, A notion of nonlocal curvature. Numer. Funct. Anal. Optim. **35**, 793–815 (2014)
2. L. Acuña Valverde, Heat content estimates over sets of finite perimeter. J. Math. Anal. Appl. **441**, 104–120 (2016)
3. L. Acuña Valverde, Heat content for stable processes in domains of $\mathbb{R}^d$. J. Geom. Anal. **27**, 492–524 (2017)
4. F. Alter, V. Caselles, Uniqueness of the Cheeger set of a convex body. Nonlinear Anal. Theory Methods Appl. **70**, 32–44 (2009)
5. F. Alter, V. Caselles, A. Chambolle, A characterization of convex calibrable sets in $\mathbb{R}^N$. Math. Ann. **332**, 329–366 (2005)
6. L. Ambrosio, N. Fusco, D. Pallara, *Functions of Bounded Variation and Free Discontinuity Problems*. Oxford Mathematical Monographs (Clarendon, Oxford, 2000)
7. L. Ambrosio, G. De Philippis, L. Martinazzi, Gamma-convergence of nonlocal perimeter functionals. Manuscripta Math. **134**, 377–403 (2011)
8. L. Ambrosio, J. Bourgain, H. Brezis, A. Figalli, BMO-type norms related to the perimeter of sets. Commun. Pure Appl. Math. **69**(6), 1062–1086 (2016)
9. F. Andreu, V. Caselles, J.M. Mazón, *Parabolic Quasilinear Equations Minimizing Linear Growth Functionals*. Progress in Mathematics, vol. 223 (Birkhauser, Basel, 2004)
10. F. Andreu, J.M. Mazón, J.D. Rossi, J. Toledo, A nonlocal p-Laplacian evolution equation with nonhomogeneous Dirichlet boundary conditions. SIAM J. Math. Anal. **40**, 1815–1851 (2008/2009)
11. F. Andreu, J.M. Mazón, J.D. Rossi, J. Toledo, A nonlocal p-Laplacian evolution equation with Neumann boundary conditions. J. Math. Pures Appl. **90**, 201–227 (2008)
12. F. Andreu-Vaillo, J.M. Mazón, J.D. Rossi, J. Toledo, *Nonlocal Diffusion Problems*. Mathematical Surveys and Monographs, vol. 165 (American Mathematical Society, Providence, 2010)
13. P. Arias, V. Caselles, G. Facciolo, V. Lazcano, R. Sadek, Nonlocal variational models for inpainting and interpolation. Math. Models Methods Appl. Sci. **22**(Suppl. 2), 1230003, 65 pp. (2012)
14. J.-F. Aujol, G. Gilboa, N. Papadakis, Fundamentals of non-local total variation spectral theory, in *Proceedings of the Scale and Variational Methods in Computer Vision*, pp. 66–77 (2015)
15. A. Baernstein, Integral means, univalent functions and circular symmetrization. Acta Math. **133**, 139–169 (1974)

© Springer Nature Switzerland AG 2019
J. M. Mazón et al., *Nonlocal Perimeter, Curvature and Minimal Surfaces for Measurable Sets*, Frontiers in Mathematics,
https://doi.org/10.1007/978-3-030-06243-9

16. P. Bénilan, M.G. Crandall, Completely accretive operators, in *Semigroups Theory and Evolution Equations (Delft, 1989)*, ed. by P. Clement et al. Lecture Notes in Pure and Applied Mathematics, vol. 135 (Marcel Dekker, New York, 1991), pp. 41–75

17. J. Bourgain, H. Brezis, P. Mironescu, *Another Look at Sobolev Spaces*, ed. by J.L. Menaldi et al. Optimal Control and Partial Differential Equations. A volume in honour of A. Bensoussan's 60th birthday (IOS Press, Amsterdam, 2001), pp. 439–455

18. L. Brasco, E. Lindgren, E. Parini, The fractional Cheeger problem. Interfaces Free Bound. **16**, 419–458 (2014)

19. H. Brezis, *Operateurs Maximaux Monotones* (North Holland, Amsterdam, 1973)

20. H. Brezis, How to recognize constant functions. Usp. Mat. Nauk **57** (2002) (in Russian). English translation in Russian Math. Surveys **57**, 693–708 (2002)

21. H. Brezis, *Functional Analysis, Sobolev Spaces and Partial Differential Equations* (Universitext/Springer, Heidelberg, 2011)

22. H. Brezis, New approximations of the total variation and filters in imaging. Atti Accad. Naz. Lincei Rend. Lincei Mat. Appl. **26**(2), 223–240 (2015)

23. H. Brezis, H.-M. Nguyen, Two subtle convex nonlocal approximations of the BV-norm. Nonlinear Anal. **137**, 222–245 (2016)

24. A. Burchard, Cases of equality in the Riesz rearrangement inequality. Ann. Math. (2) **143**(3), 499–527 (1996)

25. X. Cabré, E. Cinti, Energy estimates and 1-D symmetry for nonlinear equations involving the half-Laplacian. Discrete Contin. Dyn. Syst. **28**(3), 1179–1206 (2010)

26. X. Cabré, E. Cinti, Sharp energy estimates for nonlinear fractional diffusion equations. Calc. Var. Partial Differ. Equ. **49**(1–2), 233–269 (2014)

27. X. Cabré, Y. Sire, Nonlinear equations for fractional Laplacians, I: regularity, maximum principles, and Hamiltonian estimates. Ann. Inst. H. Poincaré Anal. Non Lineaire **31**(1), 23–53 (2014)

28. X. Cabré, Y. Sire, Nonlinear equations for fractional Laplacians II: existence, uniqueness, and qualitative properties of solutions. Trans. Am. Math. Soc. **367**(2), 911–941 (2015)

29. L. Caffarelli, J.M. Roquejoffre, O. Savin, Nonlocal minimal surfaces. Commun. Pure Appl. Math. **63**, 1111–1144 (2010)

30. L. Caffarelli, P.E. Souganidis, Convergence of nonlocal threshold dynamics approximations to front propagation. Arch. Ration. Mech. Anal. **195**, 1–23 (2010)

31. L. Cafarelli, E. Valdinoci, Uniform estimates and limiting arguments for nonlocal minimal surfaces. Calc. Var. Partial Differ. Equ. **41**, 203–240 (2011)

32. A. Chambolle, M. Morini, M. Ponsiglione, Nonlocal curvature flows. Arch. Ration. Mech. Anal. **218**, 1263–1329 (2015)

33. E. Cinti, J. Serra, E. Valdinoci, Quantitative flatness results and BV–estimates for stable nonlocal minimal surfaces. J. Differ. Geom. (to appear)

34. M. Cozzi, A. Figalli, Regularity theory for local and nonlocal minimal surfaces: an overview, in *Nonlocal and Nonlinear Diffusions and Interactions: New Methods and Directions*. Lecture Notes in Mathematics, 2186, CIME/CIME Foundation Subseries (Springer, Cham, 2017), pp. 117–158

35. W. Cygan, T. Grzywny, Heat content for convolution semigroups. J. Math. Anal. Appl. **446**, 1393–1414 (2017)

36. J. Dávila, On an open question about functions of bounded variation. Calc. Var. Partial Differ. Equ. **15**, 519–527 (2002)

37. J. Davila, M. del Pino, J. Wei, Nonlocal s-minimal surfaces and Lawson cones. J. Differ. Geom. **109**(1), 111–175 (2018)

38. E. De Giorgi, Definizione ed espressione analitica del perimetro di un insieme. Atti Accad. Naz. Lincei. Rend. Cl. Sci. Fis. Mat. Nat. **14**, 390–393 (1953)
39. E. De Giorgi, Su una teoria generale della misura $(r - 1)$-dimensionale in uno spazio ad r dimensioni. (Italian) Ann. Mat. Pura Appl. **36**, 191–213 (1954)
40. I. Ekeland, *Convexity Methods in Hamiltonian Mechanics* (Springer, Berlin, 1990)
41. A. El Chakik, A. Elmoataz, X. Desquesnes, Mean curvature flow on graphs for image and manifold restoration and enhancement. Signal Process. **105**, 449–463 (2014)
42. H. Federer, *Geometric Measure Theory*. Die Grundlehren der mathematischen Wissenschaften, Band 153 (Springer, New York 1969)
43. H. Federer, W. Fleming, Normal and integral currents. Ann. Math. **72**, 458–520 (1960)
44. W. Feller, *An Introduction to Probability Theory and Its Applications*, vol. II, 2nd edn. (Wiley, New York, 1971), pp. xxiv+669
45. A. Figalli, N. Fusco, F. Maggi, V. Millot, M. Morini, Isoperimetric and stability properties of balls with respect to nonlocal energies. Commun. Math. Phys. **336**, 441–507 (2015)
46. H.W. Fleming, R. Rishel, An integral formula for total gradient variation. Arch. Math. **11**, 218–222 (1960)
47. R.L. Frank, R. Seiringer, Non-linear ground state representations and sharp Hardy inequalities. J. Funct. Anal. **255**, 3407–3430 (2008)
48. G. Franzina, E. Valdinoci, Geometric analysis of fractional phase transition interfaces, in *Geometric Properties for Parabolic and Elliptic PDEs*. Springer INdAM Series, vol. 2 (Springer, Milano, 2013), pp. 117–130
49. V. Fridman, B. Kawohl, Isoperimetric estimates for the first eigenvalue of the p-Laplace operator and the Cheeger constant. Comment. Math. Univ. Carol. **44**, 659–667 (2003)
50. G. Gilboa, S. Osher, Nonlocal operators with applications to image processing. SIAM Multiscale Model. Simul. **7**, 1005–1028 (2008)
51. E. Giusti, *Minimal Surface and Functions of Bounded Variation*. Monographs in Mathematics, vol. 80 (Birkhäuser, Basel, 1984)
52. D. Grieser, The first eigenvalue of the Laplacian, isoperimetric constants, and the max ow min cut theorem. Arch. Math. (Basel) **87**, 75–85 (2006)
53. Y. Hafiene, J. Fadili, A. Elmoataz, Nonlocal p-Laplacian evolution problems on graphs. SIAM J. Numer. Anal. **56**, 1064–1090 (2018)
54. G.H. Hardy, J.E. Littlewood, G. Polya, Inequalities (Cambridge University Press, Cambridge, 1952)
55. M. Ledoux, Semigroup proofs of the isoperimetric inequality in Euclidean and Gauss space. Bull. Sci. Math. **118**, 485–510 (1994)
56. E.H. Lieb, M. Loss, *Analysis*. AMS Graduate Studies in Mathematics, vol. 14 (American Mathematical Society, Providence, 1987)
57. F. Maggi, *Sets of Finite Perimeter and Geometric Variational Problems: An Introduction to Geometric Measure Theory*. Cambridge Studies in Advanced Mathematics (Cambridge University Press, Cambridge, 2012)
58. J.M. Mazón, J.D. Rossi, J. Toledo, *Nonlocal Perimeter, Curvature and Minimal Surfaces for Measurable Sets*. J. d'Anal. Math. (to appear)
59. J.M. Mazón, J.D. Rossi, J. Toledo, The heat content for nonlocal diffusion with non-singular kernels. Adv. Nonlinear Stud. **17**(2), 255–268 (2017)
60. M. Miranda, Distribuzioni aventi derivate misure insiemi di perimetro localmente finito. Ann. Scuola Norm. Sup. Pisa **18**, 27–56 (1964)
61. M. Miranda Jr., D. Pallara, F. Paronetto, M. Preunkert, Short-time heat flow and functions of bounded variation in $\mathbb{R}^N$. Ann. Fac. Sci. Toulouse **16**, 125–145 (2007)

62. A. Ponce, A new approach to Sobolev spaces and connections to Γ-convergence. Calc. Var. Partial Differ. Equ. **19**, 229–255 (2004)
63. A. Ponce, An estimate in the spirit of Poincare's inequality. J. Eur. Math. Soc. **6**, 1–15 (2004)
64. M. Preunkert, A Semigroup version of the isoperimetric inequality. Semigroup Forum **68**, 233–245 (2004)
65. M. Saez, E. Valdinoci, On the evolution by fractional mean curvature. arXiv: 1511.06944v2
66. O. Savin, E. Valdinoci, Regularity of nonlocal minimal cones in dimension 2. Calc. Var. Partial Differ. Equ. **48**, 33–39 (2013)
67. Y. Sire, E. Valdinoci, Fractional Laplacian phase transitions and boundary reactions: a geometric inequality and a symmetry result. J. Funct. Anal. **256**(6), 1842–1864 (2009)
68. E. Valdinoci, A fractional framework for perimeter and phase transitions. Milan J. Math. **81**, 1–23 (2013)
69. M. van der Berg, Heat flow and perimeter in $\mathbb{R}^m$. Potential Anal. **39**, 369–387 (2013)
70. M. van der Berg, K. Gitting, Uniform bounds for the heat content of open set in Euclidean spaces. Diff. Geom. Appl. **40**, 67–85 (2015)
71. M. van der Berg, J.F. Le Gall, Mean curvature and the heat equation. Math. Z. **215**, 437–464 (1994)
72. J. Van Schaftingen, M. Willem, *Set Transformations, Symmetrizations and Isoperimetric Inequalities.* (English summary) Nonlinear Analysis and Applications to Physical Sciences (Springer Italia, Milan, 2004), pp. 135–152
73. A. Visintin, Nonconvex functionals related to multiphase systems. SIAM J. Math. Anal. **21**, 1281–1304 (1990)
74. A. Visintin, Generalized coarea formula and fractal sets. Jpn. J. Indust. Appl. Math. **8**, 175–201 (1991)

Index

© Springer Nature Switzerland AG 2019
J. M. Mazón et al., *Nonlocal Perimeter, Curvature and Minimal Surfaces for Measurable Sets*, Frontiers in Mathematics,
https://doi.org/10.1007/978-3-030-06243-9

Printed in the United States
By Bookmasters